MACHINERY'S HANDBOOK GUIDE

23rd Edition Companion

This book should be used
in conjunction with the
twenty-third edition of
MACHINERY'S HANDBOOK

MACHINERY'S HANDBOOK

GUIDE

TO THE USE OF TABLES AND FORMULAS

23rd Edition Companion

Hundreds of Examples and Test Questions on the Use of Tables, Formulas, and General Data in Machinery's Handbook, Selected Especially for Engineering and Trade Schools, Apprenticeship and Home-Study Courses, to Ensure the Most Effective Use of the Handbook and a Thorough Knowledge of its Contents

By John M. Amiss, Franklin D. Jones, and Henry H. Ryffel

INDUSTRIAL PRESS INC.

Copyright 1931, 1939, 1951, 1954, © 1959,
© 1964, © 1968, © 1971, © 1975, © 1980, © 1984, © 1988
INDUSTRIAL PRESS INC.
200 Madison Avenue
New York, N.Y. 10016-4078

Library of Congress Cataloging-in-Publication Data

Amiss, John Milton, 1887-
  Machinery's handbook guide: hundreds of examples and
test questions on the use of tables, formulas, and general data
in Machinery's handbook . . . / by John M. Amiss, Franklin
D. Jones, and Henry H. Ryffel.
        p.   cm.
  ''23rd edition companion.''
  This book should be used in conjunction with the twenty-
third edition of Machinery's handbook.''
  Includes index.
  ISBN 0-8311-1201-8
  1. Mechanical engineering—Handbooks, manuals, etc.   I.
Jones, Franklin Day, 1879-   . II. Ryffel, Henry H.   III.
Title.   IV. Title: Machinery's handbook.
TJ151.A448   1988   87-38146
621.8′0212—dc19   CIP

Composition by David E. Seham Associates, Inc.,
Metuchen, N.J.
Printed and bound by Arcata Graphics, Fairfield, Pa.

Printed in the United States of America

4   6   8   7   5

CONTENTS

Section		Page
1	Dimensions and Areas of Circles	1
2	Chordal Dimensions, Segments, and Spheres	4
3	Formulas and Their Rearrangement	8
4	Calculations Involving Logarithms of Numbers	22
5	Dimensions, Areas and Volumes of Geometrical Figures	24
6	Geometrical Propositions and Constructions	28
7	Functions of Angles and Use of Tables	32
8	Solution of Right-angle Triangles	40
9	Solution of Oblique Triangles	61
10	Figuring Tapers	71
11	Tolerances and Allowances for Machine Parts	76
12	Using Standards Data and Information	90
13	Standard Screw and Pipe Threads	95
14	Problems in Mechanics	104
15	Strength of Materials	120
16	Design of Shafts and Keys for Power Transmission	132
17	Splines	141

Contents

18 Problems in Designing and Cutting
     Gears   149

19 Cutting Speeds, Feeds, and Machining
     Power   174

20 Numerical Control   182

21 General Review Questions   188

22 Answers to Problems and Questions   197

   Index   225

THE PURPOSE OF THIS BOOK

An engineering handbook is an essential part of the equipment of practically all engineers, machine designers, draftsmen, tool engineers, and skilled mechanics in machine shops and toolrooms. The daily use of such a book, with its various tables and general data, saves a lot of time and labor. But to obtain the full value of any handbook, the user must know enough about the contents to apply the tables, formulas, and other data, whenever they can be used to advantage.

One purpose of this book, which is based on MACHINERY'S HANDBOOK, is to show by examples, solutions, and test questions, typical applications of handbook information in both drafting rooms and machine shops. Another function is to familiarize engineering students or other users with the HANDBOOK'S contents. A third objective is to provide test questions and drill work that will enable the HANDBOOK user, through practice, to obtain the required information quickly.

MACHINERY'S HANDBOOK, in common with all other handbooks, presents information in condensed form so that a large variety of subjects can be covered in a single volume. Because of this condensed treatment, any engineering handbook must be primarily a work of reference rather than a textbook, and the practical application of some parts will not always be apparent especially to those who have had little experience in engineering work. The questions and examples in this book are intended not only to supplement some of the HANDBOOK material, but to stimulate interest both in those parts that are used frequently and in the more special sections that may be very valuable even though seldom required.

A thorough working knowledge of an engineering handbook is certain to prove valuable to everyone engaged in the design or manufacture of mechanical equipment, and this collection of problems and questions is intended to assist students and the younger engineers in acquiring such knowledge.

INTRODUCTION TO THE METRIC SYSTEM

Those who have purchased the more recent editions of MACHINERY'S HANDBOOK will have noted that a considerable amount of metric material in terms of text, tables, and formulas has been added beginning with the 19th Edition. This material was added because a large part of the world has gone "metric" and the movement in that direction continues in all countries, including the United States, that intend to compete in the international marketplace.

Thus, it seemed advisable to meet the needs of MACHINERY'S HANDBOOK users who had already started or were soon going to use the metric system.

The approved metric system in use throughout the world and one that is being introduced into the United States is called Système International, or SI for short.

An explanation of SI metric system will be found on HANDBOOK pages 2411–2413. A brief history is given of the development of this system and a description is provided for each of its seven basic units. Also shown are factors and prefixes for forming decimal multiples and submultiples of the SI units.

On pages 2415 and 2416 are given two tables of SI units. The first table covers the basic units, gives their symbols and definitions, and covers SI units having special names. The second table lists SI units with complex names and provides symbols for them.

The reader is invited next to consider the tables of conversion factors which extend from page 2417 through 2424. These tables provide factors for converting English units to metric units, or vice versa, and cover units of length, area, volume (including capacity), velocity, acceleration, flow, mass, density, force, force per unit length, bending moment or torque, moment of inertia, section modulus, momentum, pressure, stress, energy, work, power, and viscosity. Using the factors in these tables it is a simple matter of multiplication to convert from one system of units to the other. It will be noted that where the conversion factors are exact they are given to only 3 or 4 significant figures, but where they are not exact they are given to 7 significant figures to permit the maximum degree of accuracy ordinarily required in the metalworking field to be obtained.

To avoid the necessity for the use of some of the conversion factors, various conversion tables are given beginning on page 2424 and extending through page 2444. The tables for length conversion on pages 2425 to 2429 will probably be the most frequently used. It may be noted that two different types of tables are shown. The one on page 2425 facilitates converting lengths up to 100 inches into millimeters, in steps of a ten-thousandth of an inch; and up to 1000 millimeters to inches, in steps of a thousandth of a millimeter.

To make possible such a wide range in a compact table the reader must usually take two or more numbers from the table and add them together as is explained in the accompanying text. The table on pages 2426 to 2429, however, has a much more limited range of conversion for inches to millimeters and millimeters to inches. However, this table has the advantage of being direct reading; that is, only a single value is taken from the table and no addition is required.

For those who are engaged in design work where it is necessary to do computations in the fields of mechanics and strength of materials, a considerable amount of guidance will be found for the use of metric units. Thus, beginning on page 129, the use of the metric SI system in mechanics calculations is explained in detail. In succeeding pages, boldface type is used to highlight references to metric units in the Mechanics section. Metric formulas are provided also, to parallel the formulas for English units.

This same procedure is followed in the Strength of Materials section. Thus, on page 209, it is explained in boldface type that SI metric units can be applied in the calculations in place of the English units of measurement without changes to the formulas for simple stresses.

The reader should also be aware that certain tables in the HANDBOOK, such as that on page 70 which gives segments of circles for a radius = 1, can be used for either English or metric units as is indicated directly under the table heading. There are other instances, however, where separate tables are needed, such as are shown on pages 1005 to 1008 for the conversion of revolutions per minute into cutting speed in feet per minute on pages 1005 and 1006, and into cutting speed in meters per minute on pages 1007 and 1008.

It is believed that the additional metric material provided in the 19th Edition, together with that added in subsequent Editions, will provide considerable useful data and assistance to engineers and personnel in the shop who are required to use metric units of measurements. It is strongly suggested that all readers, whether or not they are using metric units at the present time, become familiar with the SI system by reading the explanatory material in the HANDBOOK and by studying the SI units and the ways of converting English units to them.

Henry H. Ryffel
Editor

CAUTION TO THE READER

This guide contains numerous references to subjects in MACHINERY'S HANDBOOK. The references to the handbook page numbers apply to the 23rd Edition of MACHINERY'S HANDBOOK.

DIMENSIONS AND AREAS OF CIRCLES

Circumferences of circles are used in calculating speeds of rotating machine parts, including: drills, reamers, cutters, grinding wheels, gears and pulleys. These speeds are variously referred to as: surface speed, circumferential speed, and peripheral speed; meaning in each case the distance a point on the surface or circumference would travel per minute. This distance is usually expressed as feet per minute. Circumferences are also required in calculating the circular pitch of gears, laying out involute curves, finding the lengths of arcs, and in solving many geometrical problems. Letters from the Greek alphabet are frequently used to designate angles, and the Greek letter π (pi) is always used to indicate the ratio between the circumference and the diameter of a circle:

$$\pi = 3.14159265 \ldots = \frac{\text{circumference of circle}}{\text{diameter of circle}}$$

For most practical purposes the value of $\pi = 3.1416$ may be used.

Example 1:—Find the circumference and area of a circle whose diameter is 8 inches.

On Handbook page 54 the circumference C of a circle is given as $3.1416d$. Therefore, $3.1416 \times 8 = 25.1328$ inches.

On the same page, the area is given as $0.7854d^2$. Therefore, A (area) $= 0.7854 \times 8^2 = 0.7854 \times 64 = 50.2656$ square inches.

Example 2:—From page 60 of the Handbook the area of a cylindrical surface equals $S = 3.1416 \times d \times h$. For a diameter of 8 inches and a height of 10 inches, the area is $3.1416 \times 8 \times 10 = 251.328$ square inches.

Example 3:—For the cylinder in Example 2 but with the area of both ends included, the total area is the sum of the area found in Example 2 plus two times the area found in Example 1. Thus, $251.328 + 2 \times 50.2656 = 351.8592$ square inches. The same result could have been obtained by using the formula for total area given on Handbook page 60: $A = 3.1416 \times d \times (\frac{1}{2}d + h) = 3.1416 \times 8 \times (\frac{1}{2} \times 8 + 10) = 351.8592$ square inches.

Example 4:—If the circumference of a tree is 96 inches, what is its diameter? Since the circumference of a circle $C = 3.1416 \times d$, $96 = 3.1416 \times d$ so that $d = 96 \div 3.1416 = 30.558$ inches.

Example 5:—The table on page 1005 of the Handbook provides values of revolutions per minute required to produce various cutting speeds for workpieces of selected diameters. How are these speeds calculated? Cutting speed in feet per minute is calculated by multiplying the circumference in feet of a workpiece by the rpm of the spindle: cutting speed in fpm = circumference in feet × rpm. Transposing this formula as explained in Section 3,

$$\text{rpm} = \frac{\text{cutting speed, fpm}}{\text{circumference in feet}}$$

For a 3-inch diameter workpiece (¼-foot diameter) and for a cutting speed of 40 fpm, rpm = $40 \div (3.1416 \times \frac{1}{4}) = 50.92 = 51$ rpm, approximately, which is the same as the value given on page 1005 of the Handbook.

PRACTICE EXERCISES FOR SECTION 1

For answers to all practice exercise problems or questions see Section 22

1. Find the area of a circle 10 mm in diameter. Its circumference.

2. On Handbook page 1007, for a 5 mm diameter rotating at 318 rpm the corresponding cutting speed is given as 5 meters per minute. Check this value.

3. For a cylinder 100 mm in diameter and 10 mm high, what is the surface area not including the top or bottom?

4. A steel column carrying a load of 10,000 pounds has a diameter of 10 inches. What is the pressure on the floor in pounds per square inch?

5. What is the ratio of the area of a square of any size to the area of a circle having the same diameter as one side of the square?

6. What is the ratio of the area of a circle to the area of a square inscribed in that circle?

7. The drilling speed for cast iron is assumed to be 70 feet per minute. Find the time required to drill two holes in each of 500 castings if each hole has a diameter of ¾ inch and is 1 inch deep. Use 0.010 inch feed and allow one-fourth minute per hole for set-up.

8. Find the weight of a cast-iron column 10 inches in diameter and 10 feet high. Cast iron weighs 0.26 pound per cubic inch.

9. If machine steel has a tensile strength of 55,000 pounds per square inch, what should be the diameter of a rod to support 36,000 pounds if the safe working stress is assumed to be one-fifth of the tensile strength?

10. Moving the circumference of a 16-inch automobile flywheel two inches, moves the camshaft through how many degrees? (The camshaft rotates at one-half the flywheel speed.)

11. The tables beginning on page 962 give lengths of chords for spacing off circumferences of circles into equal parts. Is another method available?

SECTION 2

CHORDAL DIMENSIONS, SEGMENTS, AND SPHERES

HANDBOOK Pages 62 and 962–964

A chord of a circle is the distance along a straight line from one point on the circumference to any other point. A segment of a circle is that part or area between a chord and the arc it intercepts. The lengths of chords and the dimensions and areas of segments are often required in mechanical work.

Lengths of Chords.—The table of chords, Handbook page 962, can be applied to a circle of any diameter as explained and illustrated by examples on that page. This table is given to six decimal places so that it can be used in connection with precision tool work.

Example 1:—A circle has 56 equal divisions and the chordal distance from one division to the next is 2.156 inches. What is the diameter of the circle?

The chordal length in the table for 56 divisions and a diameter of 1 equals 0.05607; therefore, in this case

$$2.156 = 0.05607 \times \text{diameter}$$

$$\text{Diameter} = \frac{2.156}{0.05607} = 38.452 \text{ inches}$$

Example 2:—A drill jig is to have 8 holes equally spaced around a circle 6 inches in diameter. How can the chordal distance between adjacent holes be determined when the table, Handbook page 962, is not available?

One-half the angle between the radial center-lines of adjacent holes = 180 ÷ number of holes. If the sine of this angle is multiplied by the diameter of the circle, the product equals the chordal distance. In this example we have 180 ÷ 8 = 22.5 degrees. The sine of 22.5 degrees (see page 104) is 0.38268; hence, the chordal distance = 0.38268 × 6 = 2.296 inches. The result is the same as would be

4

obtained with the table on Handbook page 962 because the figures in the column "Length of Chord" represent the sines of angles equivalent to 180 divided by the different numbers of spaces.

Use of the Table of Segments of Circles—Handbook pages 70 and 71.—This table is of the unit type in that the values all apply to a radius of 1. As explained above the table, the value for any other radius can be obtained by multiplying the figures in the table by the given radius, except in the case of areas when the *square* of the given radius is used. Thus, the unit type of table is universal in its application.

Example 3:—Find the area of a segment of a circle, the center angle of which is 57 degrees and the radius 2½ inches.

First locate 57 degrees in the center angle column; opposite this figure in the area column will be found 0.07808. Since the area is required, this number is multiplied by the square of 2½. Thus,

$$0.07808 \times (2½)^2 = 0.488 \text{ square inch}$$

Example 4:—A cylindrical oil tank is 4½ feet in diameter, 10 feet long, and is in a horizontal position. When the depth of the oil is 3 feet 8 inches, what is the number of gallons of oil?

The total capacity of the tank equals $0.7854 \times 4½^2 \times 10 = 159$ cubic feet.

One U. S. gallon equals 0.1337 cubic foot (see Handbook page 2404); hence, the total capacity of the tank equals $159 \div 0.1337 = 1190$ gallons.

The unfilled area at the top of the tank is a segment having a height of 10 inches or $^{10}/_{27}$ (0.37037) of the tank radius. The nearest decimal equivalent to $^{10}/_{27}$ in Column h of the table on pages 70 and 71 is 0.3707; hence, the number of cubic feet in the segment-shaped space = $(27^2 \times 0.401 \times 120) \div 1728 = 20.3$ cubic feet and $20.3 \div 0.1337 = 152$ gallons. Therefore, when the depth of oil is 3 feet 8 inches, there are $1190 - 152 = 1038$ gallons. (See also Handbook page 48 for additional information on the capacity of cylindrical tanks.)

Spheres—Handbook page 62.—Handbook page 62 gives formulas for calculating spherical volumes.

Example 5:—If the diameter of a sphere is 24⅝ inches what is the volume, given the formula:

$$\text{Volume} = 0.5236 \, d^3$$

The cube of 24⅝ = 14932.369; hence, the volume of this sphere = 0.5236 × 14932.369 = 7818.5 cubic inches.

PRACTICE EXERCISES FOR SECTION 2

For answers to all practice exercise problems or questions
see Section 22

1. Find the lengths of chords when the number of divisions of a circumference and the radii are as follows: 30 and 4; 14 and 2½; 18 and 3½.

2. Find the chordal distance between the graduations for thousandths on the following dial indicators: (*a*) Starrett has 100 divisions and 1⅜-inch dial. (*b*) Brown & Sharpe has 100 divisions and 1¾-inch dial. (*c*) Ames has 50 divisions and 1⅝-inch dial.

3. The teeth of gears are evenly spaced on the pitch circumference. In making a drawing of a gear, how wide should the dividers be set to space 28 teeth on a 3-inch diameter pitch circle?

4. In a drill jig, 8 holes, each ½ inch diameter, were spaced evenly on a 6-inch diameter circle. To test the accuracy of the jig, plugs were placed in adjacent holes and the distance over the plugs was measured with a micrometer. What should be the micrometer reading?

5. In the preceding problem, what should be the distance over plugs placed in alternate holes?

6. What is the length of the arc of contact of a belt over a pulley 2 feet 3 inches in diameter if the arc of contact is 215 degrees?

7. Find the areas, length of chords and heights, of the following segments: (*a*) radius 2 inches, angle 45 degrees; (*b*) radius 6 inches, angle 27 degrees.

8. Find the number of gallons of oil in a tank 6 feet in diameter and 12 feet long if the tank is in a horizontal position and the oil measures 2 feet deep.

9. Find the surface area of the following spheres, the diameters of which are: 1½; 3⅜; 65; 20¾.

10. Find the volume of each sphere in the above exercise.

11. The volume of a sphere is 1,802,725 cubic inches. What is its surface area and diameter?

FORMULAS AND THEIR REARRANGEMENT

A formula may be defined as a mathematical rule expressed by signs and symbols instead of in actual words. In formulas, letters are used to represent numbers or *quantities,* the term "quantity" being used to designate any number involved in a mathematical process. The use of letters in formulas, in place of the actual numbers, simplifies the solution of problems, and makes it possible to condense into small space the information that otherwise would be imparted by long and cumbersome rules. The figures or values for a given problem are inserted in the formula according to the requirements in each specific case. When the values are thus inserted, in place of the letters, the result or answer is obtained by ordinary arithmetical methods. There are two reasons why a formula is preferable to a rule expressed in words. 1. The formula is more concise, it occupies less space, and it is possible to see at a glance, the whole meaning of the rule laid down. 2. It is easier to remember a brief formula than a long rule, and it is, therefore, of greater value and convenience.

Example 1:—In spur gears, the outside diameter of the gear can be found by adding 2 to the number of teeth, and dividing the sum obtained by the diametral pitch of the gear. This rule can be expressed very simply by a formula. Assume that we write D for the outside diameter of the gear, N for the number of teeth, and P for the diametral pitch. Then the formula would be:

$$D = \frac{N + 2}{P}$$

This formula reads exactly as the rule given above. It says that the outside diameter (D) of the gear equals 2 added to the number of teeth (N), and this sum is divided by the pitch (P).

If the number of teeth in a gear is 16 and the diametral pitch 6,

8

then simply put these figures in the place of N and P in the formula, and find the outside diameter as in ordinary arithmetic.

$$D = \frac{16 + 2}{6} = \frac{18}{6} = 3 \text{ inches}$$

Example 2:—The formula for the horsepower of a steam engine is as follows:

$$H = \frac{P \times L \times A \times N}{33,000}$$

in which H = indicated horsepower of engine;

$\quad\quad\quad P$ = mean effective pressure on piston in pounds per square inch;

$\quad\quad\quad L$ = length of piston stroke in feet;

$\quad\quad\quad A$ = area of piston in square inches;

$\quad\quad\quad N$ = number of strokes of piston per minute.

Assume that $P = 90, L = 2, A = 320$, and $N = 110$; what would be the horsepower?

If we insert the given values in the formula, we have:

$$H = \frac{90 \times 2 \times 320 \times 110}{33,000} = 192$$

From the examples given, we may formulate the following general rule: *In formulas, each letter stands for a certain dimension or quantity; when using a formula for solving a problem, replace the letters in the formula by the figures given for a certain problem, and find the required answer as in ordinary arithmetic.*

Omitting Multiplication Signs in Formulas.—In formulas, the sign for multiplication ($\times$) is often left out between letters the values of which are to be multiplied. Thus AB means $A \times B$, and the formula

$$H = \frac{P \times L \times A \times N}{33,000} \text{ can also be written } H = \frac{PLAN}{33,000}$$

If $A = 3$, and $B = 5$, then: $AB = A \times B = 3 \times 5 = 15$.

It is only the multiplication sign ($\times$) that can be thus left out between the symbols or letters in a formula. All other signs must be indicated the same as in arithmetic. The multiplication sign can never be left out between two figures: 35 always means thirty-five, and "three times five" must be written 3 × 5; but "three times A" may be written 3A. As a general rule the figure in an expression such as "3A" is written first, and is known as the *coefficient* of A. If the letter is written first, the multiplication sign is not left out, but the expression is written "A × 3."

Rearrangement of Formulas.—A formula can be rearranged or "transposed" to determine the values represented by different letters of the formula. To illustrate by a simple example, the formula for determining the speed (s) of a driven pulley when its diameter (d), and the diameter (D) and speed (S) of the driving pulley are known, is as follows: $s = \dfrac{S \times D}{d}$. If the speed of the driven pulley is known and the problem is to find its diameter or the value of d instead of s, this formula can be rearranged or changed. Thus: $d = \dfrac{S \times D}{s}$.

Rearranging a formula in this way is governed by four general rules.

Rule 1. An independent term preceded by a plus sign (+) may be transposed to the other side of the equals sign (=) if the plus sign is changed to a minus sign (−).

Rule 2. An independent term preceded by a minus sign may be transposed to the other side of the equals sign if the minus sign is changed to a plus sign.

As an illustration of these rules, if A = B − C, then C = B − A, and if A = C + D − B, then B = C + D − A. That the foregoing is correct may be proved by substituting numerical values for the different letters and then transposing them as shown.

Rule 3. A term which multiplies all the other terms on one side of the equals sign may be moved to the other side, if it is made to divide all the terms on that side.

As an illustration of this rule, if A = BCD, then $\dfrac{A}{BC} = D$ or according to the common arrangement $D = \dfrac{A}{BC}$. Suppose, in the preceding formula, that B = 10, C = 5, and D = 3; then A = 10 × 5 × 3 = 150, and $\dfrac{150}{10 \times 5} = 3$.

Rule 4. A term which divides all the other terms on one side of the equals sign may be moved to the other side, if it is made to multiply all the terms on that side.

To illustrate, if $s = \dfrac{SD}{d}$, then $sd = SD$, and, according to Rule 3 $d = \dfrac{SD}{s}$. This formula may also be rearranged for determining the values of S and D; thus $\dfrac{ds}{D} = S$, and $\dfrac{ds}{S} = D$.

If, in the rearrangement of formulas, minus signs precede quantities, the signs may be changed to obtain positive rather than minus quantities. All the signs on both sides of the equals sign or on both sides of the equation may be changed. For example, if $-2A = -B + C$, then $2A = B - C$. The same result would be obtained by placing all the terms on the opposite side of the equals sign which involves changing signs. For instance, if $-2A = -B + C$, then $B - C = 2A$.

Fundamental Laws Governing Rearrangement.—After a few fundamental laws which govern any formula or equation are understood, its solution usually is very simple. An equation states that one quantity equals another quantity. So long as both parts of the equation are treated exactly alike the values remain equal. Thus, in the equation $A = \frac{1}{2}\,ab$, which states that the area A of a triangle equals one half the product of the base a times the altitude b, each side of the equation would remain equal if we added the same amount. $A + 6 = \frac{1}{2}\,ab + 6$; or we could subtract an equal amount from both sides: $A - 8 = \frac{1}{2}\,ab - 8$; or multiply both parts by the same number: $7\,A = 7\,(\frac{1}{2}\,ab)$; or we could divide both parts by the same number and we would still have a true equation.

One formula for the total area T of a cylinder is: $T = 2\pi r^2 + 2\pi rh$, where r = the radius and h = the height of the cylinder. Suppose we want to solve this equation for h. $2\pi rh + 2\pi r^2 = T$. Transposing the part which does not contain h to the other side by changing its sign, we get: $2\pi rh = T - 2\pi r^2$. In order to obtain h, we can divide both sides of the equation by any quantity which will leave h on the left-hand side thus:

$$\frac{2\pi rh}{2\pi r} = \frac{T - 2\pi r^2}{2\pi r}$$

It is clear that in the left-hand member, the $2\pi r$ will cancel out, leaving: $h = \dfrac{T - 2\pi r^2}{2\pi r}$. The expression $2\pi r$ in the right hand member cannot be cancelled because it is not an independent factor since the numerator equals the difference between T and $2\pi r^2$.

Example 3:—Rearrange the formula for a trapezoid (Handbook page 52) to obtain h.

$$A = \frac{(a + b)h}{2}$$

$2A = (a + b)h$ (multiply both members by 2)

$(a + b)h = 2A$ (transpose both members so as to get the multiple of h on the left-hand side)

$\dfrac{(a + b)h}{a + b} = \dfrac{2A}{a + b}$ (divide both members by $a + b$)

$h = \dfrac{2A}{a + b}$ (cancel $a + b$ from the left-hand member)

Example 4:—The formula for determining the radius of a sphere (Handbook page 62) is as follows:

$$r = \sqrt[3]{\frac{3V}{4\pi}}$$

Rearrange to obtain a formula for finding the volume V.

$r^3 = \dfrac{3V}{4\pi}$ (cube each side)

$4\pi r^3 = 3V$ (multiply each side by 4π)

$3V = 4\pi r^3$ (transpose both members)

$\dfrac{3V}{3} = \dfrac{4\pi r^3}{3}$ (divide each side by 3)

$V = \dfrac{4\pi r^3}{3}$ (cancel 3 from left-hand member)

The procedure has been shown in detail to indicate the underlying principles involved. The rearrangement could be simplified somewhat by direct application of the rules previously given. To illustrate:

$$r^3 = \frac{3V}{4\pi} \quad \text{(cube each side)}$$

$$4\pi r^3 = 3V \quad \text{(applying \textit{Rule 4} move } 4\pi \text{ to left-hand side)}$$

$$\frac{4\pi r^3}{3} = V \quad \text{(move 3 to left-hand side—\textit{Rule 3})}$$

This final equation would, of course, be reversed to locate V at the left of the equals sign as this is the usual position for whatever letter represents the quantity or value to be determined.

Example 5:—It is required to determine the diameter of cylinder and length of stroke of a steam engine to deliver 150 horsepower. The mean effective steam pressure is 75 pounds; the number of strokes per minute is 120. The length of the stroke is to be 1.4 times the diameter of the cylinder.

First insert in the horsepower formula (Example 2) the known values:

$$150 = \frac{75 \times L \times A \times 120}{33,000} = \frac{3 \times L \times A}{11}$$

The last expression is found by cancellation.

Assume now that the diameter of the cylinder in inches equals D. Then $L = \frac{1.4D}{12} = 0.117D$, according to the requirements in the problem; the divisor 12 is introduced to change the inches to feet, L being in feet in the horsepower formula. The area $A = D^2 \times 0.7854$. If we insert these values in the last expression in our formula, we have:

$$150 = \frac{3 \times 0.117D \times 0.7854D^2}{11} = \frac{0.2757D^3}{11}$$

$$0.2757D^3 = 150 \times 11 = 1650$$

$$D^3 = \frac{1650}{0.2757} \quad D = \sqrt[3]{\frac{1650}{0.2757}} = \sqrt[3]{5984.8} = 18.15$$

Hence, diameter of the cylinder should be about 18¼ inches, and the length of the stroke 18.15 × 1.4 = 25.41, or, say, 25½ inches.

Solving Equations or Formulas by Trial.—One of the equations used for spiral gear calculations, when the shafts are at right angles, the ratios are unequal, and the center distance must be exact, is as follows:

$$R \sec \alpha + \operatorname{cosec} \alpha = \frac{2CP_n}{n}$$

In this equation

R = ratio of number of teeth in large gear to number in small gear;

C = exact center distance:

P_n = normal diametral pitch;

n = number of teeth in small gear.

The exact spiral angle α of the large gear is found by trial using the equation just given.

Equations of this form are solved by trial by selecting an angle assumed to be approximately correct, and inserting the secant and cosecant of this angle in the equation, adding the values thus obtained, and comparing the sum with the known value to the right of the equals sign in the equation. An example will show this more clearly. Using the problem given in MACHINERY'S HANDBOOK (page 1916) as an example, R = 3; C = 10; P_n = 8; n = 28.

Hence, the whole expression

$$\frac{2CP_n}{n} = \frac{2 \times 10 \times 8}{28} = 5.714$$

from which it follows that:

$$R \sec \alpha + \operatorname{cosec} \alpha = 5.714$$

In the problem given, the approximate spiral angle required is 45 degrees. The spiral gears, however, would not meet all the conditions given in the problem, if the angle could not be slightly modified. In order to determine whether the angle should be greater or smaller than 45 degrees, insert the values of the secant and cosecant of 45 degrees in the formula. The secant of 45 degrees is 1.4142, and the cosecant, 1.4142. Then,

$$3 \times 1.4142 + 1.4142 = 5.6568$$

The value 5.6568 is too small, as it is less than 5.714, which is the required value. Hence, try 46 degrees. The secant of 46 degrees is 1.4395, and the cosecant, 1.3902. Then,

$$3 \times 1.4395 + 1.3902 = 5.7087$$

Obviously an angle of 46 degrees is too small. Proceed, therefore, to try an angle of 46 degrees, 30 minutes. This angle will be found too great. Similarly 46 degrees, 15 minutes, if tried, will be found too great, and by repeated trials it will finally be found that an angle of 46 degrees, 6 minutes, the secant of which is 1.4422, and the co-secant, 1.3878, meets the requirements. Then,

$$3 \times 1.4422 + 1.3878 = 5.7144$$

which is as close to the required value as necessary.

In general, when an equation must be solved by the trial-and-error method, all the known quantities may be written on the right-hand side of the equal sign, and all the unknown quantities on the left-hand side. A value is assumed for the unknown quantity. This value is substituted in the equation, and all the values thus obtained on the left-hand side are added. In general, if the result is greater than the known values on the right-hand side, the assumed value of the unknown quantity is too great. If the result obtained is smaller than the sum of the known values, the assumed value for the unknown quantity is too small. By thus adjusting the value of the unknown quantity until the left-hand member of the equation with the assumed value of the unknown quantity will just equal the known quantities on the right-hand side of the equal sign, the correct value of the unknown quantity may be determined.

Derivation of Formulas.—Most formulas in engineering handbooks are given without showing how they have been derived or originated, because engineers and designers usually want only the final results; moreover, such derivations would require considerable additional space and they belong in textbooks rather than in handbooks which are primarily works of reference. Although MACHINERY'S HANDBOOK contains thousands of standard and special formulas, it is apparent that no handbook can include every kind of formula, because a great many formulas apply only to local designing or manufacturing problems. Such special formulas are derived by engineers and designers

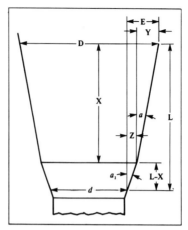

Fig. 1.   To Find Dimension X from a Given Diameter D to the Intersection of Two Conical Surfaces

for their own use. The exact methods of deriving formulas are based upon mathematical principles as they are related to the particular factors in each case. A few examples will be given to show how several different types of special formulas have been derived.

Example 6:—The problem is to deduce the general formula for finding the point of intersection of two tapers with reference to measured diameters on those tapers. In the diagram, Fig. 1,

L = the distance between the two measured diameters, D and d;

X = the required distance from one measured diameter to the intersection of tapers;

a = angle of long taper as measured from center line;

a_1 = angle of short taper as measured from center line.

Then

$$E = \frac{D - d}{2} = Z + Y$$

$$Z = (L - X) \tan a_1$$

$$Y = X \tan a$$

Therefore:

$$\frac{D - d}{2} = (L - X) \tan a_1 + X \tan a$$

and

$$D - d = 2 \tan a_1 (L - X) + 2X \tan a \qquad (1)$$

But

$$2 \tan a_1 = T_1 \qquad \text{and} \qquad 2 \tan a = T$$

in which T and T_1 represent the long and short tapers per inch, respectively.

Therefore from Equation (1)

$$D - d = T_1(L - X) + TX$$

$$D - d = T_1L - T_1X + TX$$

$$X(T_1 - T) = T_1L - (D - d)$$

$$X = \frac{T_1L - (D - d)}{T_1 - T}$$

Example 7:—A flywheel is 16 feet in diameter (outside measurement) and the center of its shaft is 3 feet above the floor. Derive a formula for determining how long the hole in the floor must be to permit the flywheel to turn.

The conditions are as represented in Fig. 2. The line AB is the

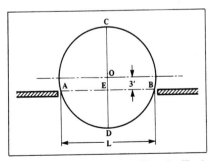

Fig. 2. To Find Length of Hole in Floor for Flywheel

floor level and is a chord of the arc ADB; it is parallel to the horizontal diameter through the center O. CD is the vertical diameter and is perpendicular to AB. It is shown in geometry that the diameter CD bisects the chord AB at the point of intersection E. Now, one of the most useful theorems of geometry is that when a diameter bisects a chord, the product of the two parts of the diameter is equal to the square of one half the chord; in other words, $AE^2 = ED \times EC$. If AB is represented by L and OE by a, $ED = r - a$ and $EC = r + a$, in which $r =$ the radius OC; hence,

$$\left(\frac{L}{2}\right)^2 = (r - a)(r + a) = r^2 - a^2$$

$$\frac{L}{2} = \sqrt{r^2 - a^2}$$

and

$$L = 2\sqrt{r^2 - a^2}$$

Substituting the values given,

$$L = 2\sqrt{8^2 - 3^2} = 14.8324 \text{ feet} = 14 \text{ feet, 10 inches}$$

The length of the hole should be at least 15 feet, to allow for clearance.

Empirical Formulas.—Many formulas used in engineering calculations cannot be established fully by mathematical derivation, but must be based upon actual tests instead of relying upon mere theories or assumptions which might introduce excessive errors. These formulas are known as "empirical formulas." Usually such a formula contains a constant (or constants in some cases) which represents the result of the tests; consequently the value obtained by the formula is consistent with these tests or with actual practice.

A simple example of an empirical formula will be found on Handbook page 384. This particular formula contains the constant 54,000 which was established by tests, and the formula is used to obtain the breaking load of wrought-iron crane chains to which a factor of safety of 3, 4, or 5 is then applied to obtain the working load. Other examples of empirical formulas will be found on page 288.

On the Handbook page 304 is an example of an empirical formula based upon experiments made with power-transmitting shafts. This formula gives the diameter of shaft required to prevent excessive twisting of shafts.

Parentheses.—Two important rules relating to the use of parentheses are based upon the principles of positive and negative numbers:

1. If a parenthesis is preceded by a + sign, it may be removed, if the terms within the parentheses retain their signs.

$$a + (b - c) = a + b - c$$

2. If a parenthesis is preceded by a − sign, it may be removed, if the signs preceding each of the terms inside of the parentheses are changed (+ changed to −, and − to +). Multiplication and division signs are not affected.

$$a - (b - c) \quad = a - b + c$$
$$a - (- b + c) = a + b - c$$

Knowledge of algebra is not necessary in order to make successful use of formulas of the general type such as are found in engineering handbooks; it is only necessary to thoroughly understand the use of letters or symbols in place of numbers, and to be well versed in the methods, rules, and processes of ordinary arithmetic. Knowledge of algebra becomes necessary only in cases where a general rule or formula which gives the answer to a problem directly is not available. In other words, algebra is useful in *developing* or originating a general rule or formula, but the formula can be *used* without recourse to algebraic processes.

Constants.—A constant is a value that does not change or is not variable. However, constants at one stage of a mathematical investigation may be variables at another stage, but an *absolute constant* has the same value under all circumstances. The ratio of the circumference to the diameter of a circle, or 3.1416, is a simple example of an absolute constant. In a common formula used for determining the indicated horsepower of a reciprocating steam engine, the product of the mean effective pressure, the length of the stroke in feet, the area of the piston in square inches, and the number of piston strokes per minute is divided by the constant 33,000, which represents the number of foot pounds of work per minute equivalent to one horsepower. Constants occur in many mathematical formulas.

Mathematical Signs and Abbreviations—Handbook page 1.—Every division of mathematics has its traditions, customs and signs, which are frequently of ancient origin. Hence we encounter Greek letters in many problems where it would seem that English letters would do as well or better. Most of the signs on page 1 will be used frequently. They should therefore be understood.

Conversion Tables.—In some cases it may be necessary to convert English units of measurement into metric units and vice versa. The tables provided in the latter portion of the Handbook will be found useful in this connection.

PRACTICE EXERCISES FOR SECTION 3
For answers to all practice exercise problems or questions
see Section 22

1. An approximate formula for determining the horsepower H of automobile engines is: $H = D^2SN/3$, where D = diameter of bore, inches; S = length of stroke, inches; N = number of cylinders. Find the horsepower of the following automobile engine: (a) bore, 3½ inches; stroke, 4½ inches; 6 cylinders. (b) Using the reciprocal of 3, how could this formula be stated?

2. Using the right-angle triangle formula: $C = \sqrt{a^2 + b^2}$ where a = one side, b = the other side and C = the hypotenuse, find the hypotenuse of a right triangle whose sides are 16 inches and 63 inches.

3. A formula for finding the blank diameter of a cylindrical shell is: $D = \sqrt{d \times (d + 4\,h)}$, where D = blank diameter; d = diameter of the shell; h = height of the shell. Find the diameter of the blank to form a cylindrical shell 3 inches diameter and 2 inches high.

4. If D = diagonal of a cube; d = diagonal of face of a cube; s = side of a cube and V = volume of a cube; then $d = \sqrt{\dfrac{2D^2}{3}}$;

$s = \sqrt{\dfrac{D^2}{3}}$ and $V = s^3$. Find the side, volume of a cube, and diagonal of the face of a cube if the diagonal of the cube is 10.

5. The area of an equilateral triangle equals one fourth of the side squared times the square root of 3, or $A = \dfrac{S^2}{4}\sqrt{3} = 0.43301S^2$.

Find the area of an equilateral triangle the side of which is 14.5 inches.

6. The formula for the volume of a sphere is: $\dfrac{4\pi r^3}{3}$ or $\dfrac{\pi d^3}{6}$. What constants may be used in place of $\dfrac{4\pi}{3}$ and $\dfrac{\pi}{6}$?

7. The formula for the volume of a solid ring is $2\pi^2 Rr^2$, where r = radius of cross section and R = radius from the center of the ring to the center of the cross section. Find the volume of a solid ring made from 2-inch round stock if the mean diameter of the ring is 6 inches.

8. Explain the following signs: $\pm$, $>$, $<$, tan, $\angle$, $\sqrt[4]{\ }$, log, θ, β, $\sin^{-1} a$, $::$, I.H.P.

9. The area A of a trapezoid (see Handbook page 52) is found by the formula:

$$A = \frac{(a + b)h}{2}$$

Transpose the formula for determining width a.

10. $R = \sqrt{r^2 + \dfrac{s^2}{4}}$; solve for r.

11. $P = 3.1416 \sqrt{2(a^2 + b^2)}$; solve for a.

12. $\cos A = \sqrt{1 - \sin^2 A}$; solve for $\sin A$.

13. $\dfrac{a}{\sin A} = \dfrac{b}{\sin B}$; solve for a, b, $\sin A$, $\sin B$.

CALCULATIONS INVOLVING LOGARITHMS
OF NUMBERS

HANDBOOK Pages 28–32 and 1676–1679

The Tables of Logarithms in earlier editions of MACHINERY'S HANDBOOK were used to shorten long, tedious calculations involving multiplication, division, obtaining the powers of numbers, and extracting the roots of numbers. By using the logarithms of numbers, multiplication becomes simple addition of the logarithms of the numbers to give the logarithm of the product from which the product itself is then found. For division, subtraction of the logarithm of the divisor from that of the dividend gives the logarithm of the quotient from which the quotient may be found.

The availability of inexpensive handheld calculators has eliminated the need to use logarithms for such purposes; there are, however, many applications in which the logarithm of a number is used in the same way that a trigonometric function such as sine or tangent is used. For example, at the top of page 29 there is a formula to find the number of years n that money is lent. The example accompanying the formula shows the necessary calculations that include the use of the logarithms 3, 2.69897, and 0.025306 which correspond to the numbers 1000, 500, and 1.06, respectively. These logarithms were obtained directly from a handheld calculator and are the common or *Briggs* system logarithms which have a base of 10. Any other system of logarithms such as those of base e ($e = 2.71828 . . .$) could have been used in this problem with the same result. Base e logarithms are sometimes referred to as "natural logarithms."

There are, however, other kinds of problems in which logarithms of a specific base, usually 10 or e must be used to obtain the correct result. On the logarithm keys of most calculators the base 10 logs are identified by the word "log" while those of base e are referred to as "ln."

As an example of logarithms that are part of a formula or equation, the formula for the mean effective pressure of a steam cylinder is calculated from the formula:

$$p = P (1 + \ln R)/R$$

If $P = 120$ and $R = 2.89$, then, from a calculator, $\ln 2.89 = 1.06125$ so that $p = 120 \times (1 + 1.06125)/2.89 = 85.59$. Note that in this formula the logarithm to base e was required. If by error the logarithm of 2.89 for base 10, namely, $\log 2.89 = 0.46090$ had been used, the result would have been $p = 120 \times (1 + 0.46090)/2.89 = 60.66$ which is erroneous.

In some instances logarithms may be used to expedite problems in which more than one solution is possible and several choices must be compared. The Logarithms of Gear Ratios described on page 1676 of the Handbook, together with the tables beginning on page 1679 represent one such instance. Here, the logarithms of gear ratios provide the means to obtain several possible solutions, of varying degrees of accuracy, to a change-gear ratio problem.

PRACTICE EXERCISES FOR SECTION 4

For answers to all practice exercise problems or questions
see Section 22

1. Why has the use of logarithms to simplify multiplication and division been discontinued?

2. Is there a procedure to calculate the most accurate set of change gears possible without using logarithms?

3. Is the base e of the "natural" logarithm system ever used directly in a calculation?

SECTION 5

DIMENSIONS, AREAS AND VOLUMES OF GEOMETRICAL FIGURES

HANDBOOK Pages 46–65; 69–73

The formulas given for the solution of different problems relating to the areas of surfaces and volumes of various geometrical figures are derived from plane and solid geometry. For purposes of shop mathematics, all that is necessary is to select the appropriate figure and use the formula given. Keep in mind the tables which have been studied and use them in the solution of the formulas whenever this can be done to advantage.

Many rules may be developed directly from the table for polygons on Handbook page 72. These rules will enable one to solve very easily nearly every problem involving a regular polygon. For instance, in the first "A" columns at the left, $A = 7.6942 \, S^2$ for a decagon; hence in this case $S = \sqrt{A \div 7.6942}$. In the first "R" column, $R = 1.3066 \, S$ for an octagon; hence $S = R \div 1.3066$.

The frequent occurrence of such geometrical figures as squares, hexagons, spheres and spherical segments in shop calculations causes the tables dealing with these figures to be very useful.

Example 1:—A rectangle 12 inches long has an area of 120 square inches; what is the length of its diagonal?

The area of a rectangle equals the product of the two sides; hence, the unknown side of this rectangle equals $^{120}/_{12} = 10$ inches.

Length of diagonal $= \sqrt{12^2 + 10^2} = \sqrt{244} = 15.6205$.

Example 2:—If the diameter of a sphere, the diameter of the base, and the height of a cone are all equal, find the volume of the sphere if the volume of the cone is 250 cubic inches.

The formula on Handbook page 50 for the volume of a cone, shows that in this case $250 = 0.2618 \, d^2 h$ in which $d =$ diameter of cone base and $h =$ vertical height of cone; hence,

$$d^2 = \frac{250}{0.2618 \, h}$$

24

Since in this example d and h are equal,

$$d^2 = \frac{250}{0.2618}$$

and

$$d = \sqrt[3]{\frac{250}{0.2618}} = 9.8474 \text{ inches}$$

Referring to the formula on Handbook page 62, the volume of a sphere = $0.5236\ d^3 = 0.5236 \times 9.8474^3 = 500$ cubic inches.

In solving the following exercises, first, construct the figure carefully and then apply the formula. Use the examples in the Handbook as models.

PRACTICE EXERCISES FOR SECTION 5

For answers to all practice exercise problems or questions
see Section 22

1. Find the volume of a cylinder having a base radius of 12.5 and a height of 16.3.

2. Find the area of a triangle the sides of which are 12, 14 and 18 inches in length.

3. Find the volume of a torus or circular ring made from 1½-inch round stock, if its outside diameter is 14 inches.

4. A bar of hexagonal screw stock measures 0.750 inch per side. What is the largest diameter that can be turned from this bar?

5. Using the prismoidal formula (Handbook page 46), find the volume of the frustum of a regular triangular pyramid if its lower base is 6 inches per side; upper base 2 inches per side and height 3 inches. (Use table on page 72 for areas. The side of the midsection equals one-half the sum of one side of the lower base and one side of the upper base.)

6. What is the diameter of a circle the area of which is equivalent to that of a spherical zone whose radius is 4 inches and height 2 inches?

7. Find the volume of a steel ball ⅜ inches in diameter.

8. What is the length of the side of a cube if the volume equals

the volume of a frustum of a pyramid with square bases, 4 inches and 6 inches per side and 3 inches high?

9. Find the volume of a bronze bushing if its inside diameter is 1 inch, outside diameter is 1½ inches, and length is 2 inches.

10. Find the volume of material making up a hollow sphere if its outside diameter is 10 inches, inside diameter 6 inches.

11. Find the area of a polygon of 10 sides, inscribed in a 6-inch circle.

12. What is the radius of a fillet if its chord is 2 inches? What is its area?

13. Find the area of the conical surface and volume of a frustum of a cone if the diameter of its lower base is 3 feet, diameter of upper base 1 foot and height 3 feet.

14. Find the total area of the sides and the volume of a triangular prism 10 feet high, having a base width of 8 feet.

15. The diagonal of a square is 16 inches. What is the length of its side?

16. How many gallons can be contained in a barrel having the following dimensions: height 2½ feet; bottom diameter 18 inches; bilge diameter 21 inches? (The sides are formed to the arc of a circle.)

17. Find the area of a sector of circle if the radius is 8 inches and central angle is 32 degrees.

18. Find the height of a cone if its volume is 17.29 cubic inches and the radius of its base is 4 inches.

19. Find the volume of a rectangular pyramid having a base 4 × 5 inches and height 6 inches.

20. Find the distance across the corners of both hexagons and squares when the distance across flats in each case is: ½; 1⅝; 3³⁄₁₀; 5; 8.

21. The diagonals of squares are: 2.0329; 4.6846. Find the length of each side.

22. In measuring the distance over plugs in a die which has six ¾-inch holes equally spaced on a circle, what should be the mi-

crometer reading over opposite plugs, if the distance over alternate plugs is 4½ inches?

23. To what diameter should a shaft be turned in order to mill on one end a hexagon 2 inches on a side; an octagon 2 inches on a side?

GEOMETRICAL PROPOSITIONS AND
CONSTRUCTIONS

HANDBOOK Pages 36–45

Geometry is that branch of mathematics which deals with the relations of lines, angles, surfaces, and solids. Plane geometry treats of lines, angles, and surfaces in one plane only; and, since this branch of geometry is of especial importance in mechanical work, the various propositions or fundamental principles are given in the Handbook, also various problems or constructions. This information is particularly useful in mechanical drafting and in solving problems in mensuration.

Example 1:—A segment-shaped casting (see illustration) has a chordal length of 12 inches and the height of the chord is two inches; determine by the application of a geometrical principle the radius *R* of the segment.

This problem may be solved by the application of the first geometrical proposition given on Handbook page 40. In this example one chord consists of two sections *a* and *b*, each 6 inches long; the other intersecting chord consists of one section *d*, 2 inches long, and the length of section *c* is to be determined in order to find radius *R*. Since $a \times b = c \times d$, it follows that

$$c = \frac{a \times b}{d} = \frac{6 \times 6}{2} = 18 \text{ inches}$$

therefore, $R = \frac{c + d}{2} = \frac{18 + 2}{2} = 10 \text{ inches}$

In this example one chordal dimension, $c + d =$ the diameter; but, the geometrical principle given in the Handbook applies regardless of the relative lengths of the intersecting chords.

Example 2:—The center lines of three holes in a jig plate form a triangle. The angle between two of these intersecting center lines is

28

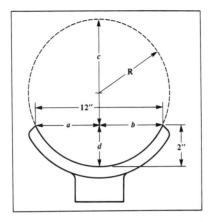

52 degrees. Another angle between adjacent center lines is 63 degrees. What is the third angle?

This problem is solved by application of the first geometrical principle on Handbook page 35. The unknown angle = 180 − (63 + 52) = 65 degrees.

Example 3:—The center lines of four holes in a jig plate form a four-sided figure. Three of the angles between the different intersecting center lines are 63 degrees, 105 degrees, and 58 degrees, respectively. What is the fourth angle?

According to the geometrical principle at the bottom of Handbook page 37, the unknown angle = 360 − (63 + 105 + 58) = 134 degrees.

Example 4:—The centers of three holes are located on a circle. The angle between the radial center lines of the first and second holes is 22 degrees, and the center-to-center distance measured along the circle is 2½ inches. The angle between the second and third holes is 44 degrees. What is the center-to-center distance along the circle?

This problem is solved by application of the fourth principle on Handbook page 40. Since the lengths of the arcs are proportional to the angles, the center distance between the second and third holes

$= \dfrac{44 \times 2\frac{1}{2}}{22} = 5$ inches. (See also rules governing proportion on

Handbook page 26.)

The following practice exercises relate to the propositions and constructions given and should be answered without the aid of the Handbook.

PRACTICE EXERCISES FOR SECTION 6

For answers to all practice exercise problems or questions
see Section 22

1. If any two angles of a triangle are known, how can the third angle be determined?

2. State three cases where one triangle is equal to another.

3. When are triangles similar?

4. What is the purpose of proving triangles similar?

5. If a triangle is equilateral, what follows?

6. What are the properties of the bisector of any angle of an equilateral triangle?

7. What is an isosceles triangle?

8. How do the size of an angle and the length of a side of a triangle compare?

9. Can you draw a triangle whose sides are 5, 6, and 11 inches?

10. What is the length of the hypotenuse of a right triangle the sides of which are 12 and 16 inches?

11. What is the value of the exterior angle of any triangle?

12. What are the relations of angles formed by two intersecting lines?

13. Draw two intersecting straight lines and a circle tangent to these lines.

14. Construct a right triangle given the hypotenuse and one side.

15. When are the areas of two parallelograms equal?

16. When are the areas of two triangles equal?

17. If a radius of a circle is perpendicular to a chord, what follows?

18. What is the relation between the radius and tangent of a circle?

19. What lines pass through the point of tangency of two tangent circles?

20. What are the attributes of two tangents drawn to a circle from an external point?

21. What is the value of an angle between a tangent and a chord drawn from the point of tangency?

22. Are all angles equal, if their vertices are on the circumference of a circle and they are subtended by the same chord?

23. If two chords intersect within a circle, what is the value of the product of their respective segments?

24. How can a right angle be drawn, using a semi-circle?

25. Upon what does the length of circular arcs in the same circle depend?

26. To what are the circumferences and areas of two circles proportional?

SECTION 7

FUNCTIONS OF ANGLES AND USE OF TABLES

HANDBOOK Pages 74–126

The basis of trigonometry is proportion. If the sides of any angle are indefinitely extended and perpendiculars from various points on one side are drawn to intersect the other side, right triangles will be formed and the ratios of the respective sides and hypotenuses will be identical. If the base of the smallest triangle thus formed is 1 inch and the altitude is ½ inch (see Fig. 1) the ratios between these sides is 1 ÷ ½ = 2 or ½ ÷ 1 = ½ depending upon how the ratio is stated. If the next triangle is measured, the ratio between the base and altitude will likewise be either 2 or ½, and this will always be true for any number of triangles, if the angle remains unchanged. For example, 3 ÷ 1½ = 2 and 4½ ÷ 2¼ = 2 or 1½ ÷ 3 = ½ and 2¼ ÷ 4½ = ½. This relationship explains why rules can be developed to find the length of any side of a triangle when the angle and one side are known or to find the angle when any two sides are known. Since there are two relations between any two sides of a triangle, there can therefore be a total of six ratios with three sides. These ratios have been defined and explained in the Handbook. Refer to pages 74 and 75 and note explanations of the terms "side adjacent," "side opposite," and hypotenuse.

The abbreviations of the trigonometrical functions begin with a small letter and are not followed by periods.

Functions of Angles and Use of Trigonometrical Tables.—At the bottom of page 74 in the Handbook are given certain rules for determining the functions of angles. These rules, which should be memorized, may also be expressed as simple formulas:

$$\text{sine} = \frac{\text{side opposite}}{\text{hypotenuse}} \qquad \text{cosecant} = \frac{\text{hypotenuse}}{\text{side opposite}}$$

$$\text{cosine} = \frac{\text{side adjacent}}{\text{hypotenuse}} \qquad \text{secant} = \frac{\text{hypotenuse}}{\text{side adjacent}}$$

32

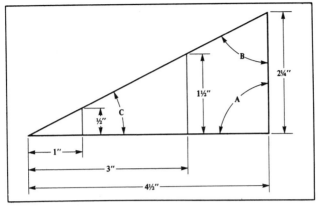

Fig. 1. For a Given Angle, the Ratio of the Base to the Altitude is the Same for All Triangle Sizes

$$\text{tangent} = \frac{\text{side opposite}}{\text{side adjacent}} \qquad \text{cotangent} = \frac{\text{side adjacent}}{\text{side opposite}}$$

Note that these functions are arranged in pairs to include sine and cosecant; cosine and secant; tangent and cotangent; and each pair consists of a function and its reciprocal. Also, note that the different functions are merely ratios, the sine being the ratio of the *side opposite* to the *hypotenuse;* cosine the ratio of the *side adjacent* to the *hypotenuse,* etc. The numbers found in the tables of trigonometric functions (Handbook pages 82 to 126) are, therefore, tables of ratios. For example, tan 20° 30′ = 0.37388; this means that in any right triangle having an acute angle of 20° 30′, the side opposite that angle is equal in length to 0.37388 times the length of the side adjacent. Cos 50° 22′ = 0.63787; this means that in any right triangle having an angle of 50° 22′, if the hypotenuse equals a certain length, say 8, the side adjacent to the angle will equal 0.63787 × 8 or 5.10296.

Referring to Fig. 1, tan angle $C = 2\frac{1}{4} \div 4\frac{1}{2} = 1\frac{1}{2} \div 3 = \frac{1}{2} \div 1 = 0.5$; therefore, for this particular angle C the *side opposite* is always equal to 0.5 times *side adjacent,* thus: $1 \times 0.5 = \frac{1}{2}$; $3 \times 0.5 = 1\frac{1}{2}$, and $4\frac{1}{2} \times 0.5 = 2\frac{1}{4}$. The side opposite angle B equals $4\frac{1}{2}$; hence, tan $B = 4\frac{1}{2} \div 2\frac{1}{4} = 2$. The use of the tables of functions will now be explained.

Finding Angle Equivalent to Given Function.—After determining the tangent of angle C or of angle B, the values of these angles can be determined readily. As tan C = 0.5, find the number nearest to this in the tangent column. On Handbook page 108 will be found 0.50004, and the corresponding angle is 26 degrees, 34 minutes. Since angle A = 90 degrees, and, as the sum of three angles of a triangle always equals 180 degrees, it is evident that angle $C + B$ = 90 degrees; therefore, B = 90 degrees minus 26 degrees, 34 minutes = 63 degrees, 26 minutes. The table on Handbook page 108 also shows that the tan 63 degrees, 26 minutes is 1.9998 or 2 within 0.0002.

Note that for angles between 45 and 90 degrees, the table is used by reading from the bottom up, and the minutes are found in the right-hand column as explained on Handbook page 81.

In the foregoing example the tangent is used to determine the unknown angles because the known sides are the *side adjacent* and the *side opposite*, these being the sides required for determining the tangent. If the side adjacent and the length of hypotenuse had been given instead, the unknown angles might have been determined by first finding the cosine because the cosine equals the side adjacent divided by the hypotenuse.

Since the acute angles (like B and C, Fig. 1) of any right triangle must be complementary, the function of any angle equals the co-function of its complement; thus, the sine of angle B = the cosine of angle C; the tangent of angle B = the cotangent of angle C, etc. Thus, tan B = $4\frac{1}{2} \div 2\frac{1}{4}$ and cotangent C also equals $4\frac{1}{2} \div 2\frac{1}{4}$. The tangent of 20° 30′ = 0.37388; which also equals the cotangent of 69° 30′. For this reason, it is only necessary to calculate the trigonometric ratios to 45° when making a table of trigonometric functions for angles between 45° and 90°, and this is why the functions of angles between 45 and 90 degrees are located in the table by reading it backwards or in reverse order, as previously mentioned.

Example 1:—Find the tangent of 44 degrees, 59 minutes.

Following instructions given on page 81 of the Handbook, find 44 degrees at the top of page 126, and 59 minutes in the left-hand column headed M; then, opposite 59 and in the column headed "Tan," find the tangent 0.99942.

Example 2:—Find the tangent of 45 degrees, 5 minutes.

The number of degrees is found at the bottom of Handbook page 126 and the number of minutes in the right-hand column headed M.

Opposite 5, and above "Tan" at the *bottom* of the table, we find the required tangent is 1.0029.

How to Find More Accurate Functions and Angles than are Given in the Table.—In engineering handbooks, the values of trigonometric functions are usually given to degrees and minutes; hence if the given angle is in degrees, minutes and seconds, the value of the function is determined from the nearest given values by interpolation.

Example 3:—Assume that the sine of 14° 22′ 26″ is to be determined. It is evident that this value lies between the sine of 14° 22′ and the sine of 14° 23′.

Sine 14° 23′ = 0.24841 and sine 14° 22′ = 0.24813. The difference = 0.24841 − 0.24813 = 0.00028. Consider this difference as a whole number (28) and multiply it by a fraction having as its numerator the number of seconds (26) in the given angle, and as its denominator 60 (number of seconds in one minute). Thus $\frac{26}{60} \times 28 = 12$ nearly; hence, by adding 0.00012 to sine of 14° 22′ we find that sine 14° 22′ 26″ = 0.24813 + 0.00012 = 0.24825.

The correction value (represented in this example by 0.00012) is *added* to the function of the *smaller* angle nearest the given angle in dealing with *sines* or *tangents* but this correction value is *subtracted* in dealing with cosines or cotangents.

Example 4:—Find the angle whose cosine is 0.27052.

The table of trigonometric functions shows that the desired angle is between 74° 18′ and 74° 19′ because the cosines of these angles are, respectively, 0.27060 and 0.27032. The difference = 0.27060 − 0.27032 = 0.00028. From the cosine of the *smaller* angle or 0.27060, subtract the given cosine; thus 0.27060 − 0.27052 = 0.00008; hence $\frac{8}{28} \times 60 = 17″$ or the number of seconds to add to the smaller angle to obtain the required angle. Thus the angle for a cosine of 0.27052 is 74° 18′ 17″. Angles corresponding to given sines, tangents, or cotangents may be determined by the same method.

Trigonometric Functions of Angles Greater than 90 Degrees.—In obtuse triangles one angle is greater than 90 degrees, and the handbook tables can be used for finding the functions of angles larger than 90 degrees.

The sine of an angle greater than 90 degrees but less than 180 degrees equals the sine of an angle which is the difference between 180 degrees and the given angle.

Example 5:—Find the sine of 118 degrees.

Sin 118° = sin (180° − 118°) = sin 62°. The sine of 118° can also be obtained directly from the table. Angle 118° will be found at the lower left-hand corner of Handbook page 110. For angles between 90 and 135 degrees, the minutes are read in the left-hand column and the functions at the bottom of the table hence, the sine 118 degrees is found in the column above "Sine" and opposite 0 in the minutes column, the sine being 0.88295. By referring to page 110, it will be seen that this is the sine given for 62 degrees.

The cosine, tangent and cotangent of an angle greater than 90 but less than 180 degrees equals, respectively, the cosine, tangent and cotangent of the difference between 180 degrees and the given angle; but in this case the angular function has a *negative* value and must be preceded by a minus sign.

Example 6:—Find tan 123 degrees, 20 minutes.

Tan 123° 20′ = −tan (180° − 123° 20′) = −tan 56° 40′ = − 1.5204. The tangent 123 degrees, 20 minutes is given directly on Handbook page 115, excepting that the minus sign must be added.

Example 7:—Find tangent 150 degrees.

For angles between 135 and 180 degrees, the minutes are found in the right-hand column and the functions at the top of the table.

On Handbook page 111, in the column headed "Tan" and opposite "0" in the right-hand minutes column, is the tangent 0.57735; and, as the value is negative, it is written −0.57735. In the calculation of triangles, it is very important to include the minus sign in connection with the cosines, tangents and cotangents of angles between 90 and 180 degrees. The handbook table on page 81 shows clearly the negative and positive values of different functions and angles.

Use of Functions for Laying Out Angles.—The tables of trigonometric functions may be used for laying out angles accurately either on drawings or in connection with templet work, etc. The following example illustrates the general method of procedure:

Example 8:—Construct or lay out an angle 27 degrees, 29 minutes by using its sine instead of a protractor.

First, draw two lines at right angles and to any convenient length. Find, in the Handbook table, the sine of 27 degrees, 29 minutes which equals 0.46149. If there is space enough, lay out the diagram to an enlarged scale in order to obtain greater accuracy. Assume, in this

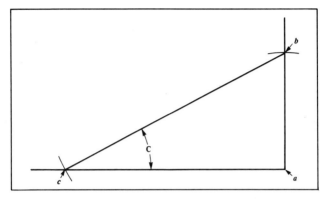

Fig. 2.   Method of Laying Out Angle by Using Its Sine

case, that the scale is to be 10 to 1: therefore, multiply the sine of the angle by 10 obtaining 4.6149 or $4\frac{39}{64}$ very nearly. Set the dividers or the compass to this dimension and with a (Fig. 2) as a center, draw an arc, thus obtaining one side of the triangle ab. Now set the compass to 10 inches (since the scale is 10 to 1) and with b as the center, describe an arc so as to obtain intersection c. The hypotenuse of the triangle is now drawn through the intersections c and b thus obtaining an angle C of 27 degrees, 29 minutes within fairly close limits. The angle C, laid out in this way, equals 27 degrees, 29 minutes because

$$\frac{\text{Side Opposite}}{\text{Hypotenuse}} = \frac{4.6149}{10} = 0.46149 = \sin 27° \; 29'$$

Table of Functions used in Conjunction with Formulas.—When milling keyways it is often desirable to know the total depth from the outside of the shaft to the bottom of the keyway. With this depth known, the cutter can be fed down to the required depth without taking any measurements other than that indicated by the graduations on the machine. In order to determine the total depth, it is necessary to calculate the height of the arc, which is designated as dimension

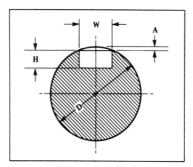

Fig. 3.   To Find Height *A* for Arc of Given Radius and Width *W*

A in Fig. 3. The formula usually employed to determine *A* for a given diameter of shaft *D* and width of key *W*, is

$$A = \frac{D}{2} - \sqrt{\left(\frac{D}{2}\right)^2 - \left(\frac{W}{2}\right)^2}$$

Another formula which is simpler than the one just given is used in conjunction with a table of trigonometric functions as arranged in MACHINERY'S HANDBOOK. The formula follows:

$$A = \frac{D}{2} \times \text{versed sine of an angle whose cosecant is } \frac{D}{W}$$

Example 9:—To illustrate the application of this formula, let it be required to find the height *A* when the shaft diameter *D* is ⅞ inch and the width *W* of the key is ⁷⁄₃₂ inch. Then,

$$\frac{D}{W} = \frac{\frac{7}{8}}{\frac{7}{32}} = \frac{7}{8} \times \frac{32}{7} = 4$$

In the table on Handbook page 96 locate the value nearest 4 in the column headed "Cosec.," which is 3.9984. Next, in the column headed "Cosine," and on the same line with this cosecant, find the value 0.96822. Subtract this value from 1 to get 0.03178 which is the versed sine according to the formula on Handbook page 81.

Then,

$$A = \frac{D}{2} \times 0.03178 = \frac{7 \times 0.03178}{8 \times 2} = 0.0139 \text{ inch}$$

The total depth of the keyway equals dimension *H* plus 0.0139 inch.

PRACTICE EXERCISES FOR SECTION 7
For answers to all practical exercise problems or questions
see Section 22

1. How should the tables be used to find angles between 45° and 90°?

2. Explain the meaning of sin 30° = 0.50000.

3. Find sin 18° 26′ 30″; tan 27° 16′ 15″; cos 32° 55′ 17″.

4. Find the angles which correspond to the following tangents; 0.52035; 0.13025; to the following cosines: 0.06826; 0.66330.

5. Give two rules for finding *side opposite* a given angle.

6. Give two rules for finding the *side adjacent* to a given angle.

7. Explain the following terms: equilateral; isosceles; acute angle; obtuse angle; oblique angle.

8. What is meant by complement; side adjacent; side opposite?

9. Can the elements just referred to be used in solving an isosceles triangle?

10. Without referring to the Handbook, show the relationship between the six trigonometric functions and an acute angle, using the terms *side opposite*, *side adjacent* and *hypotenuse* or abbreviations *SO*, *SA* and *Hyp*.

11. Construct by use of tangents an angle of 42° 20′.

12. Construct by use of sines an angle of 68° 15′.

13. Construct by use of cosines an angle of 55° 5′.

SOLUTION OF RIGHT-ANGLE TRIANGLES

HANDBOOK Page 77

A thorough knowledge of the solution of triangles or trigonometry is essential in drafting, layout work, bench work, and for convenient and rapid operation of some machine tools. Calculations concerning gears, screw-threads, dovetails, angles, tapers, solution of polygons, gage design, cams, dies, and general inspection work are dependent upon trigonometry. Many geometrical problems may be solved more rapidly by trigonometry than by geometry.

In shop trigonometry it is not necessary to develop and memorize the various rules and formulas; but it is essential that the six trigonometric functions be thoroughly mastered. It is well to remember that a thorough, working knowledge of trigonometry depends upon drill work; hence a large number of problems should be solved.

The various formulas for the solution of right-angle triangles are given on Handbook page 77 and examples showing their application on page 78. These formulas may, of course, be applied to a large variety of practical problems in drafting-rooms, tool-rooms, and machine shops, as indicated by the few examples which follow.

Whenever two sides of a right-angle triangle are given, the third side can always be found by a simple arithmetical calculation, as shown by the third and fourth examples on Handbook page 78. To find the angles, however, it is necessary to use the tables of sines, cosines, tangents and cotangents, and if only one side and one of the acute angles are given, the natural trigonometric functions must be used for finding the lengths of the other sides.

Example 1:—The Jarno taper is 0.600 inch per foot for all numbers. What is the included angle?

As the angle measured from the axis or center line is 0.600 ÷ 2 = 0.300 inch per foot, the tangent of one-half the included angle = 0.300 ÷ 12 = 0.025 = tan 1° 26'; hence the included angle = 2° 52'. A more direct method is to divide the taper per foot by 24 as explained on Handbook page 686 (see paragraph "To Find Angle for Given Taper per Foot").

Example 2:—Determine the width W (see Fig. 1) of a cutter for milling a splined shaft having 6 splines 0.312 inch wide, and a diameter B of 1.060 inches.

This dimension W may be computed by using the following formula:

$$W = \sin\left(\frac{\dfrac{360°}{N} - 2a}{2}\right) \times B$$

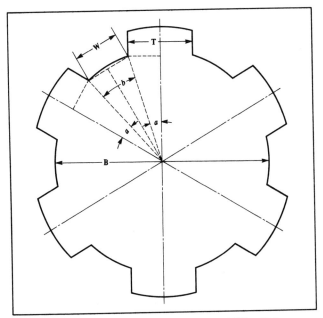

Fig. 1. To Find Width W of Spline-groove Milling Cutter

in which N = number of splines; B = diameter of body or of the shafting at the root of the splineway.

Angle a must first be computed, as follows:

$$\sin a = \frac{T}{2} \div \frac{B}{2} \text{ or } \sin a = \frac{T}{B}$$

where T = width of spline; B = diameter at the root of splineway. In this example

$$\sin a = \frac{0.312}{1.060} = 0.29434 \text{ and}$$

$$a = 17°7'; \text{ hence}$$

$$W = \sin\left(\frac{\dfrac{360°}{6} - 2 \times 17° 7'}{2}\right) \times 1.060 = 0.236 \text{ inch}$$

This formula has also been used frequently in connection with broach design, but it is capable of a more general application. If the splines are to be ground on the sides, suitable deduction must be made from dimension W to leave sufficient stock for grinding.

If the angle b is known or is first determined, then

$$W = B \times \sin \frac{b}{2}$$

As there are 6 splines in this example, angle $b = 60° - 2 a = 60° - 34° 14' = 25° 46'$; hence

$$W = 1.060 \times \sin 12°53' = 1.060 \times 0.22297 = 0.236 \text{ inch}$$

Example 3:—In sharpening the teeth of thread milling cutters, if the teeth have rake, it is necessary to position each tooth for the grinding operation so that the outside tip of the tooth is at horizontal distance x from the vertical center line of the milling cutter as shown in Fig. 2B. What must this distance x be if the outside radius to the tooth tip is r and the rake angle is to be A? What distance x off center must a $4\frac{1}{2}$-inch diameter cutter be set if the teeth are to have a 3-degree rake angle?

In Fig. 2A, it will be seen that, assuming the tooth has been properly sharpened to rake angle A, if a line is drawn extending the front edge of the tooth, it will be at a perpendicular distance x from the

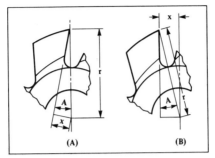

Fig. 2. To Find Horizontal Distance for Positioning Milling Cutter Tooth for Grinding Rake Angle A

center of the cutter. Let the cutter now be rotated until the tip of the tooth is at a horizontal distance x from the vertical center line of the cutter as shown in Fig. 2B. It will be noted that an extension of the front edge of the cutter is still at perpendicular distance x from the center of the cutter, indicating that the cutter face is parallel to the vertical center line or is itself vertical, which is the desired position for sharpening, using a vertical wheel. Thus, x is the proper offset distance for grinding the tooth to rake angle A if the radius to the tooth tip is r. Since r is the hypotenuse and x is one side of a right-angled triangle,

$$x = r \sin A$$

For a cutter diameter of $4\frac{1}{2}$ inches and a rake angle of 3 degrees,

$$x = (4.5 \div 2) \sin 3° = 2.25 \times 0.05234$$
$$= 0.118 \text{ inch}$$

Example 4:—Forming tools are to be made for different sizes of poppet valve heads and a general formula is required for finding angle x from dimensions given in Fig. 3.

The values for b, h, and r can be determined easily from the given dimensions. Angle x can then be found in the following manner: Referring to the lower diagram,

$$\tan A = \frac{h}{b} \quad (1) \qquad c = \frac{h}{\sin A} \quad (2)$$

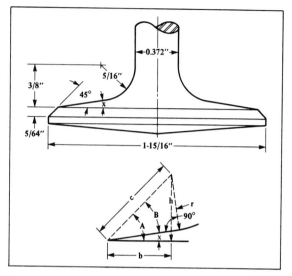

Fig. 3. To Find Angle *x*, Having the Dimensions Given on the Upper Diagram

Also,

$$c = \frac{r}{\sin B} = \frac{r}{\sin (A - x)} \qquad (3)$$

From Equations (2) and (3) by comparison,

$$\frac{r}{\sin (A - x)} = \frac{h}{\sin A}$$

$$\sin (A - x) = \frac{r \sin A}{h} \qquad (4)$$

From the dimensions given, it is obvious that $b = 0.392125$ inch, $h = 0.375$ inch, and $r = 0.3125$ inch. Substituting these values in Equations (1) and (4) and solving, angle A will be found to be 43 degrees, 43 minutes and angle $(A - x)$, to be 35 degrees, 10 minutes. By subtracting these two values, angle x will be found to equal 8 degrees, 33 minutes.

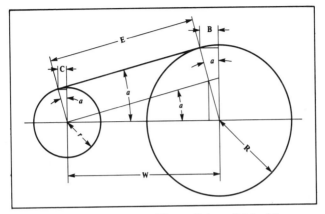

Fig. 4.   To Find Dimension *E* or Distance Between Points of Tangency

Example 5:—In tool designing it frequently becomes necessary to determine the length of a tangent to two circles. In Fig. 4, *R* = radius of large circle = $^{13}/_{16}$ inch; *r* = radius of small circle = $^3/_8$ inch; *W* = center distance between circles = $1^{11}/_{16}$ inches.

With the values given it is required to find the following: *E* = length of tangent; *B* = length of horizontal line from point of tangency on large circle to the vertical center line; and, *C* = length of horizontal line from point of tangency on small circle to the vertical center line.

$$\sin a = \frac{R - r}{W} = \frac{^{13}/_{16} - ^3/_8}{1^{11}/_{16}} = 0.25925$$

Angle *a* = $15°$ $1'$ nearly

$$E = W \cos a = 1^{11}/_{16} \times 0.9658 = 1.63 \text{ inches}$$

$$B = R \sin a \quad \text{and} \quad C = r \sin a$$

Example 6:—In a right triangle having the dimensions shown in Fig. 5, a circle is inscribed. Find the radius of the circle.

In the illustration, *BD* = *BE* and *AD* = *AF*, because "tangents drawn to a circle from the same point are equal." *EC* = *CF*, and *EC* = radius *OF*. Then let *R* = radius of inscribed circle. *AC* − *R* = *AD* and *BC* − *R* = *DB*. Adding,

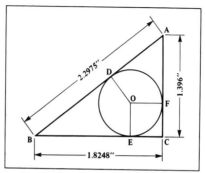

Fig. 5.   To Find Radius of Circle Inscribed in Triangle

$$AC + BC - 2R = AD + DB$$
$$AD + DB = AB$$
hence, $$AC + BC - AB = 2R$$

Stated as a rule, "The diameter of a circle inscribed in a right triangle is equal to the difference between the hypotenuse and the sum of the other sides." Substituting the given dimensions, we have $1.396 + 1.8248 - 2.2975 = 0.9233 = 2R$, and $R = 0.4616$.

Example 7:—A part is to be machined to an angle b of 30 degrees (Fig. 6), by using a vertical forming tool having a clearance angle a of 10 degrees. Calculate the angle of the forming tool as measured in a plane Z–Z which is perpendicular to the front or clearance surface of the tool.

Assume that B represents the angle in plane Z–Z.

$$\tan B = \frac{Y}{X} \quad \text{and} \quad Y = y \times \cos a \tag{1}$$

Also,

$$y = X \times \tan b \quad \text{and} \quad X = \frac{y}{\tan b} \tag{2}$$

Now substituting the values of Y and X in Equation (1), we have:

$$\tan B = \frac{y \times \cos a}{\dfrac{y}{\tan b}}$$

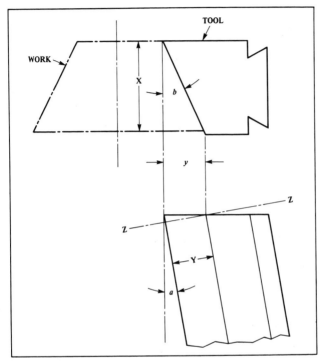

Fig. 6. The Problem is to Determine Angle of Forming Tool in Plane Z-Z

Clearing this equation of fractions,

$$\tan B = \cos a \times \tan b$$

In this example, $\tan B = 0.98481 \times 0.57735 = 0.56858$; hence $B = 29° 37'$ nearly.

Example 8:—A method of checking the diameter at the small end of a taper plug gage is shown by Fig. 7. The gage is first mounted on a sine-bar so that the top of the gage is parallel with the surface plate. A disk of known radius r is then placed in the corner formed by the end of the plug gage and the top side of the sine-bar. Now by determining the difference X in height between the top of the gage

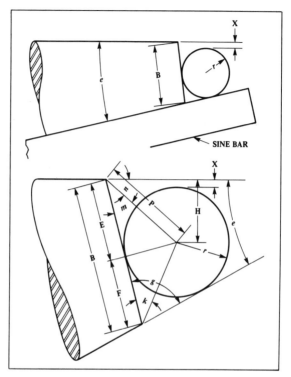

Fig. 7. The Problem is to Determine Height X in Order to Check Diameter B of Taper Plug

and the top edge of the disk, the accuracy of the diameter B can be checked readily. Derive formulas for determining dimension X.

The known dimensions are:

    e = angle of taper;
    r = radius of disk; and
    B = required diameter at end of plug gage.

$$g = 90 \text{ degrees} - \tfrac{1}{2}\,e \quad \text{and} \quad k = \tfrac{1}{2}\,g$$

By trigonometry,

$$F = \frac{r}{\tan k}; \quad E = B - F; \quad \text{and} \quad \tan m = \frac{r}{E}$$

Also

$$P = \frac{r}{\sin m}; \quad n = g - m; \quad \text{and} \quad H = P \sin n$$

Therefore, $X = H - r$ or $r - H$, depending on whether or not the top edge of the disk is above or below the top of the plug gage. In the illustration the top of the disk is below the top surface of the plug gage so that it is evident that $X = H - r$.

To illustrate the application of these formulas, assume that $e = 6$ degrees, $r = 1$ inch and $B = 2.400$ inches. The dimension X is then found as follows:

$$g = 90 - \frac{6}{2} = 87°; \quad \text{and} \quad k = 43° \; 30'$$

By trigonometry,

$$F = \frac{1}{0.94896} = 1.0538''; \quad E = 2.400 - 1.0538 = 1.3462 \text{ inches}$$

$$\tan m = \frac{1}{1.3462} = 0.74283 \quad \text{and} \quad m = 36° \; 36' \; 22''$$

$$P = \frac{1}{0.59631} = 1.6769''; \quad n = 87° - 36° \; 36' \; 22'' = 50° \; 23' \; 38''$$

and

$$H = 1.6769 \times 0.77044 = 1.2920 \text{ inches}$$

Therefore,

$$X = H - r = 1.2920 - 1 = 0.2920 \text{ inch}$$

The disk in this case is below the top surface of the plug gage, hence the formula $X = H - r$ was applied.

Example 9:—In Fig. 8, $a = 1\frac{1}{4}$ inches, $h = 4$ inches, and angle $A = 12$ degrees. Find dimension x and angle B.

Draw an arc through points E, F, and G, as shown, with r as a radius. According to a well-known theorem of geometry, which is given on page 39 of MACHINERY'S HANDBOOK, if an angle at the circumference of a circle, between two chords, is subtended by the

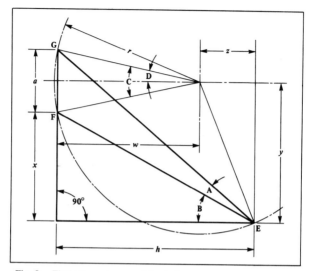

Fig. 8.   Find Dimension x and Angle B, Having a, h, and Angle A

same arc as the angle at the center, between two radii, then the angle at the circumference is equal to one-half the angle at the center. This being true, angle C is twice the magnitude of angle A, and angle D = angle A = 12 degrees. It will now be readily observed that

$$r = \frac{a}{2 \sin D} = \frac{1.25}{2 \times 0.20791} = 3.0061$$

$$w = \frac{a}{2} \cot D = 0.625 \times 4.7046 = 2.9404$$

and

$$z = h - w = 4 - 2.9404 = 1.0596$$

Now

$$y = \sqrt{r^2 - z^2} = \sqrt{7.9138505} = 2.8131$$

and

$$x = y - \frac{a}{2} = 2.8131 - 0.625 = 2.1881 \text{ inches}$$

Finally,

$$\tan B = \frac{x}{h} = \frac{2.1881}{4} = 0.54703$$

and

$$B = 28 \text{ degrees, 40 minutes, 47 seconds}$$

Example 10:—A steel ball is placed inside of a taper gage as shown in Fig. 9. If the angle of the taper, length of taper, radius of ball and its position in the gage are known, how can the end diameters X and Y of the gage be determined by measuring dimension C?

The ball should be of such size as to project above the face of the gage. This, however, is not necessary, although preferable, as it permits the required measurements to be more readily obtained. After measuring the distance C, the calculation of dimension X is as follows: First obtain dimension A, which equals R multiplied by csc a. Then adding R to A and subtracting C we obtain dimension B. Dimension X may then be obtained by multiplying 2 B by the tangent of angle a. The formulas for X and Y can therefore be written as follows:

$$X = 2(R \csc a + R - C) \tan a = 2 R \sec a + 2 \tan a(R - C)$$
$$Y = X - (2 T \tan a)$$

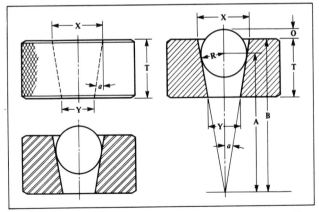

Fig. 9. Checking Dimensions X and Y by Using One Ball of Given Size

If in Fig. 9 angle a = 9 degrees, T = 1.250 inches, C = 0.250 inch and R = 0.500 inch, what are the dimensions X and Y? Applying the formula,

$$X = 2 \times 0.500 \times 1.0125 + 2 \times 0.15838 (0.500 - 0.250)$$

Solving this equation, X = 1.0917 inches. Then

$$Y = 1.0917 - (2.500 \times 0.15838) = 0.6957$$

Example 11:—In designing a motion of the type shown in Fig. 10, it is essential, usually, to have link E swing equally above and below the center line MM. A mathematical solution of this problem follows. In the illustration, G represents the machine frame; F, a lever shown in the extreme positions; E, a link; and D, a slide. The distances A and B are fixed and the problem is to obtain $A + X$, or the required length of the lever. In the right triangle:

$$A + X = \sqrt{(A - X)^2 + \left(\frac{B}{2}\right)^2}$$

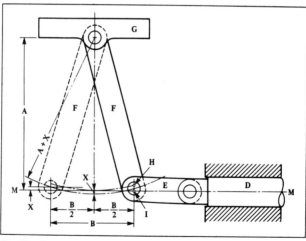

Fig. 10.   Determining Length of Link F so that Link E will Swing Equally Above and Below the Center Line

Squaring, we have:

$$A^2 + 2\,AX + X^2 = A^2 - 2\,AX + X^2 + \frac{B^2}{4}$$

$$4\,AX = \frac{B^2}{4}$$

$$X = \frac{B^2}{16\,A}$$

$$A + X = A + \frac{B^2}{16\,A} = \text{length of lever}$$

To illustrate the application of this formula, assume that the length of a lever is required when the distance $A = 10$ inches and the stroke B of the slide is 4 inches.

$$\text{Length of lever} = A + \frac{B^2}{16\,A} = 10 + \frac{16}{16 \times 10} = 10.100 \text{ inches}$$

Thus it is evident that the pin in the lower end of the lever will be 0.100 inch below the center line M–M when half the stroke has been made, and at each end of the stroke the pin will be 0.100 inch above this center line.

Example 12:—The spherical hubs of bevel gears are checked by measuring the distance x (Fig. 11) over a ball or plug placed against a plug gage which fits into the bore. Determine this distance x.

First find H by means of the formula for circular segments on Handbook page 54.

$$H = 2.531 - \tfrac{1}{2}\sqrt{4 \times 2.531^2 - 1.124^2} = 0.0632 \text{ inch}$$

$$AB = \frac{1.124}{2} + 0.25 = 0.812 \text{ inch}$$

$$BC = 2.531 + 0.25 = 2.781 \text{ inch}$$

Applying one of the formulas for right triangles, on Handbook page 77,

$$AC = \sqrt{2.781^2 - 0.812^2} = 2.6599 \text{ inches}$$

$$AD = AC - DC = 2.6599 - 2.531 = 0.1289 \text{ inch}$$

$$x = 1.094 + 0.0632 + 0.1289 + 0.25 = 1.536 \text{ inches}$$

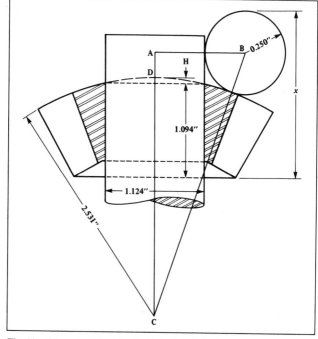

Fig. 11. Method of Checking the Spherical Hub of a Bevel Gear with Plug Gages

Example 13:— The accuracy of a gage is to be checked by placing a ball or plug between the gage jaws and measuring to the top of the ball or plug as shown by Fig. 12. Dimension *x* is required and the known dimensions and angles are shown by the illustration.

One-half of the included angle between the gage jaws equals one-half of 13° 49′ or 6° 54½′ and the latter equals angle *a*.

$$AB = \frac{0.500}{\sin 6° 54\frac{1}{2}′} = 4.1569 \text{ inches}$$

DE is perpendicular to *AB* and angle *CDE* = angle *a*; hence,

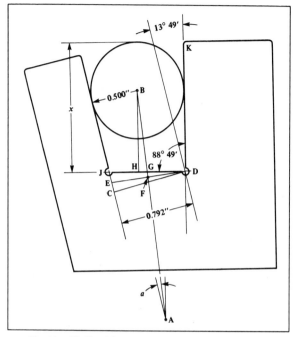

Fig. 12. Finding Dimension *x* to Check Accuracy of Gage

$$DE = \frac{CD}{\cos 6°\ 54\frac{1}{2}'} = \frac{0.792}{\cos 6°\ 54\frac{1}{2}'} = 0.79779 \text{ inch}$$

$$AF = \frac{DE}{2} \times \cot 6°\ 54\frac{1}{2}' = 3.2923 \text{ inches}$$

Angle $CDK = 90° + 132°\ 49' = 103°\ 49°$

Angle $CDJ = 103°\ 49' - 88°\ 49' = 15°$

Angle $EDJ = 15° - 6°\ 54\frac{1}{2}' = 8°\ 5\frac{1}{2}'$

$$GF = \frac{DE}{2} \times \tan 8°\ 5\frac{1}{2}' = 0.056711 \text{ inch}$$

Angle HBG = angle EDJ = 8° 5½'

$BG = AB - (GF + AF) = 0.807889$ inch

$BH = BG \times \cos 8° 5½' = 0.79984$ inch

$x = BH + 0.500 = 1.2998$ inches

If surface JD is parallel to the bottom surface of gage, the distance between these surfaces might be added to x in order to use a height gage from a surface plate.

Helix Angles of Screw Threads, Hobs, and Helical Gears.—The terms "helical" and "spiral" often are used interchangeably in drafting-rooms and shops, although the two curves are entirely different. As the illustration on Handbook page 45 shows, every point on a helix is equidistant from the axis, and the curve advances at a uniform rate around a cylindrical area. The helix is illustrated by the springs shown on Handbook page 330. A spiral is flat like a clock spring. A spiral may be defined mathematically as a curve having a constantly increasing radius of curvature.

If a piece of paper is cut in the form of a right triangle and wrapped around a cylinder, as indicated by the diagram (Fig. 13), the hypotenuse will form a helix. The curvature of a screw thread represents a helix. From the properties of a right triangle, simple formulas can be derived for determining helix angles. Thus, if the circumference of a part is divided by the lead or distance that the helix advances axially in one turn, the quotient equals the tangent of the helix angle

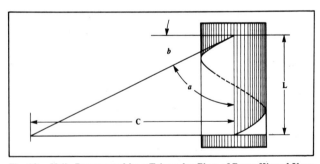

Fig. 13.   Helix Represented by a Triangular Piece of Paper Wound Upon a Cylinder

as measured from the axis. The angles of helical curves usually are measured from the axis but not invariably. The helix angle of a helical or "spiral" gear is measured from the axis but the helix angle of a screw thread is measured from a plane perpendicular to the axis. In case of a helical gear, the angle is *a* (Fig. 13), whereas, for a screw thread, the angle is *b*; hence, for helical gears tan *a* of helix angle = C/L; for screw threads, tan *b* of helix angle = L/C. The helix angle of a hob such as is used for gear cutting, also is measured as indicated at *b* and often is known as the "end angle" because it is measured from the plane of the end surface of the hob. In calculating helix angles of helical gears, screw threads, and hobs, the pitch circumference is used.

Example 14:—If the pitch diameter of a helical gear = 3.818 inches and the lead = 12 inches, what is the helix angle?

$$\tan \text{ helix angle} = \frac{3.818 \times 3.1416}{12} = 1 \text{ very nearly; hence the}$$

angle = 45 degrees.

PRACTICE EXERCISES FOR SECTION 8
For answers to all practice exercise problems or questions
see Section 22

1. The No. 4 Morse taper is 0.6233 inch per foot: calculate the included angle.

2. ANSI standard pipe threads have a taper of ¾ inch per foot. What is the angle on each side of the center line?

3. To what dimension should the dividers be set to space 8 holes evenly on a circle of 6 inches diameter?

4. Explain the derivation of the formula

$$W = \sin \left(\frac{\dfrac{360°}{N} - 2a}{2} \right) \times B$$

For notation, see Example 2 of Section 8 and the diagram Fig. 1.

5. The top of a male dovetail is 4 inches wide. If the angle is 55 degrees and the depth is ⅝ inch, what is the width at the bottom of the dovetail?

6. Angles may be laid out accurately by describing an arc with a radius of given length and then determining the length of a chord of this arc. In laying out an angle of 25 degrees, 20 minutes, using a radius of 8 inches, what should be the length of the chord opposite the angle named?

7. How large a square may be milled on the end of a 2½-inch bar of round stock?

8. A guy wire from a smoke stack is 120 feet long. How high is the stack if the wire is attached 10 feet from the top and makes an angle of 57 degrees with the stack?

9. In laying out a master jig plate, it is required that holes F and H, Fig. 14, shall be on a straight line which is $1\frac{3}{4}$ inch distant from hole E. The holes must also be on lines making, respectively, 40- and 50-degree angles with line EG, drawn at right angles to the sides of the jig plate through E, as shown in the engraving. Find the dimensions a, b, c and d.

10. In Fig. 15 is shown a template for locating a pump body on a milling fixture, the inside contour of the template corresponding with the contour of the pump flange. Find the angle a from the values given.

11. Find the chordal distances as measured over plugs placed in holes located at different radii in the taximeter drive ring shown in Fig. 16. All holes are $\frac{7}{32}$ inch diameter; the angle between the center line of each pair of holes is 60 degrees.

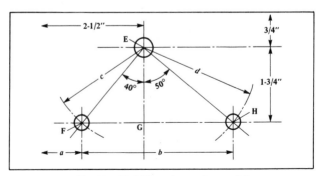

Fig. 14.   Find Dimensions a, b, c, and d

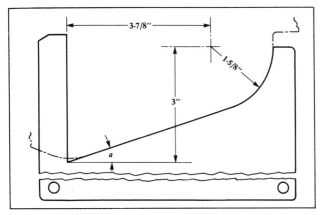

Fig. 15.   To Find Angle *a* Having the Dimensions Given

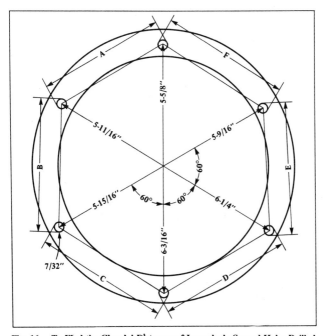

**Fig. 16.   To Find the Chordal Distances of Irregularly Spaced Holes Drilled
in a Taximeter Drive Ring**

12. An Acme screw thread has an outside diameter of $1\frac{1}{4}$ inches and has 6 threads per inch. Find the helix angle using the pitch diameter as a base. Find, also, the helix angle if a double thread is cut on the screw.

13. What is the lead of the flutes in a $\frac{7}{8}$-inch drill if the helix angle, measured from the center line of the drill, is 27° 30′?

14. A 4-inch diameter milling cutter has a lead of 68.57 inches. What is the helix angle measured from the axis?

SECTION 9

SOLUTION OF OBLIQUE TRIANGLES

HANDBOOK Page 79

In solving problems for dimensions or angles it is often convenient to work with oblique triangles. In an oblique triangle none of the angles are right angles. One of the angles may be over 90 degrees or each of the three angles may be less than 90 degrees. Any oblique triangle may be solved by constructing perpendiculars to the sides from appropriate vertices, thus forming right triangles. The methods, previously explained, for solving right triangles, will then solve the oblique triangles. The objection to this method of solving oblique triangles is that it is a long, tedious process.

Two of the examples in the Handbook on page 80, which are solved by the formulas for oblique triangles, will be solved by the right-angle triangle method. These have been solved to show that all oblique triangles can be thus solved and to give an opportunity to compare the two methods. All oblique triangles come under four classes:

(1) Given one side and two angles.
(2) Given two sides and the included angle.
(3) Given two sides and the angle opposite one of them.
(4) Given the three sides.

Example 1:—Solve the first example on Handbook page 80 by the right-angle triangle method. Referring to the accompanying Fig. 1.

$$\text{Angle } C = 180° - (62° + 80°) = 38°$$

Draw a line *DC* perpendicular to *AB*.

In the right triangle *BDC*, $\dfrac{DC}{BC} = \sin 62°$

$$\frac{DC}{5} = 0.88295; \quad DC = 5 \times 0.88295 = 4.41475$$

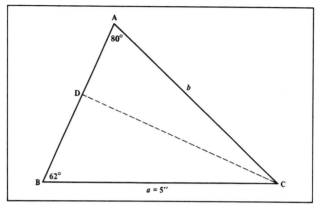

Fig. 1.   Oblique Triangle Solved by Right-angle Triangle Method

Angle $BCD = 90° - 62° = 28°$; $DCA = 38° - 28° = 10°$

$$\frac{BD}{5} = \cos 62°; BD = 5 \times 0.46947 = 2.34735$$

In triangle ADC, $\dfrac{AC}{DC} = \sec 10°$

$AC = 4.41475 \times 1.0154 = 4.4827$

$$\frac{AD}{4.41475} = \tan 10°; AD = 4.41475 \times 0.17633 = 0.7785$$
and $AB = AD + DB = 0.7785 + 2.34735 = 3.1258$
$C = 38°$; $b = 4.4828$; $c = 3.1258$

Example 2:—Apply the right-angle triangle method to the solution of the second example on Handbook page 80.

Referring to Fig. 2, draw a line BD perpendicular to CA.

In the right triangle BDC, $\dfrac{BD}{9} = \sin 35°$

$BD = 9 \times 0.57358 = 5.16222$

$$\frac{CD}{9} = \cos 35°; CD = 9 \times 0.81915 = 7.37235$$

$DA = 8 - 7.37235 = 0.62765$

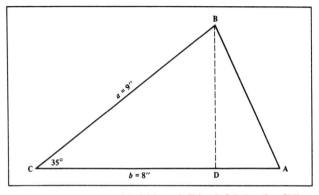

Fig. 2.   Another Example of the Right-angle Triangle Solution of an Oblique Triangle

In the right triangle BDA, $\dfrac{BD}{DA} = \dfrac{5.16222}{0.62765} = \tan A$

$\tan A = 8.2246$ and $A = 83°4'$

$B = 180° - (83°\ 4' + 35°) = 61°\ 56'$

$\dfrac{BA}{BD} = \dfrac{BA}{5.1622} = \operatorname{cosec} 83°\ 4'; BA = 5.1622 \times 1.0074 = 5.2004$

$A = 83°\ 4'; B = 61°\ 56'; C = 35°$

$a = 9; b = 8; c = 5.2004$

Use of Formulas for Oblique Triangles.—Oblique triangles are not encountered as frequently as right triangles and therefore the methods of solving the latter may be fresh in the memory while methods for solving the former may be forgotten. All the formulas involved in the solution of the four cases of oblique triangles are derived from: (1) the law of sines; (2) the law of cosines; and (3) the sum of the angles of a triangle equal 180°.

The law of sines is that in any triangle, the length of the sides are proportional to the sines of the opposite angles. (See diagrams on Handbook page 79.)

$$\frac{a}{\sin A} = \frac{b}{\sin B} = \frac{c}{\sin C} \quad (1). \text{ Solving this equation we get:}$$

$$\frac{a}{\sin A} = \frac{b}{\sin B} \quad \text{then } a \times \sin B = b \times \sin A \text{ and}$$

$$a = \frac{b \times \sin A}{\sin B}; \quad \sin B = \frac{b \times \sin A}{a}$$

$$b = \frac{a \times \sin B}{\sin A}; \quad \sin A = \frac{a \times \sin B}{b}$$

In like manner, $\frac{a}{\sin A} = \frac{c}{\sin C}$ and

$$a \times \sin C = c \times \sin A; \text{ hence } \sin A = \frac{a \times \sin C}{c}$$

and $\frac{b}{\sin B} = \frac{c}{\sin C}$ or $b \times \sin C = c \times \sin B$.

Thus twelve formulas may be derived. As a general rule only formula (1) is remembered and special formulas are derived from it as required.

The law of cosines states that in any triangle the square of any side equals the sum of the squares of the other two sides minus twice their product multiplied by the cosine of the angle between them. These relations are stated as formulas thus:

(1) $a^2 = b^2 + c^2 - 2bc \times \cos A$ or $a = \sqrt{b^2 + c^2 - 2bc \times \cos A}$

(2) $b^2 = a^2 + c^2 - 2ac \times \cos B$ or $b = \sqrt{a^2 + c^2 - 2ac \times \cos B}$

(3) $c^2 = a^2 + b^2 - 2ab \times \cos C$ or $c = \sqrt{a^2 + b^2 - 2ab \times \cos C}$

Solving (1), $a^2 = b^2 + c^2 - 2bc \times \cos A$ for $\cos A$,

$2bc \times \cos A = b^2 + c^2 - a^2$ (transposing)

$$\cos A = \frac{b^2 + c^2 - a^2}{2 bc}$$

In like manner formulas for $\cos B$ and $\cos C$ may be found.

Example 3:—A problem quite often encountered in lay-out work is illustrated in Fig. 3. It is required to find the dimensions x and y between the holes, these dimensions being measured from the intersection of the perpendicular line with the center line of the two

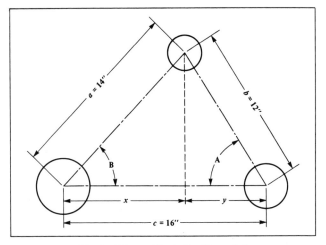

Fig. 3. Diagram Illustrating Example 3

lower holes. The three center-to-center distances are the only known values.

The method that might first suggest itself is to find the angle A (or B) by some such formula as

$$\cos A = \frac{b^2 + c^2 - a^2}{2bc} \quad (1)$$

and then solve the right triangle for y by the formula

$$y = b \cos A \quad (2)$$

Formulas (1) and (2) can be combined as follows:

$$y = \frac{b^2 + c^2 - a^2}{2c} \quad (3)$$

The value of x can be determined in a similar manner.

The second solution of this problem involves the following geometrical proposition: In any oblique triangle where the three sides are known, the ratio of the length of the base to the sum of the other two sides equals the ratio of the difference between the length of the two sides to the difference between the lengths x and y. Therefore, if $a = 14$, $b = 12$ and $c = 16$ inches, then

$$c : (a + b) = (a - b) : (x - y)$$

$$16 : 26 = 2 : (x - y)$$

$$(x - y) = \frac{26 \times 2}{16} = 3\frac{1}{4} \text{ inches}$$

$$x = \frac{(x + y) + (x - y)}{2} = \frac{16 + 3\frac{1}{4}}{2} = 9.625 \text{ inches}$$

$$y = \frac{(x + y) - (x - y)}{2} = \frac{16 - 3\frac{1}{4}}{2} = 6.375 \text{ inches}$$

When Angles Have Negative Values.—In the solution of oblique triangles having one angle larger than 90 degrees, it is sometimes necessary to use angles whose functions are negative. Review Handbook pages 15 and 81. Notice that for angles between 90 degrees and 180 degrees the cosine, tangent, cotangent and secant are negative.

Example 4:—Referring to Fig. 4, two sides and the angle between them are known. Find angles A and B. (See Handbook page 79.)

$$\tan A = \frac{4 \times \sin 20^\circ}{3 - 4 \times \cos 20^\circ} = \frac{4 \times 0.34202}{3 - 4 \times 0.93969} = \frac{1.36808}{3 - 3.75876}$$

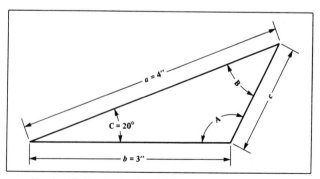

Fig. 4. Finding Angles A and B from the Dimensions Given

It will be seen that in the denominator of the fraction above, the number to be subtracted from 3 is greater than 3; the numbers are therefore reversed, 3 being subtracted from 3.75876, the remainder then being negative. Hence:

$$\tan A = \frac{1.36808}{3 - 3.75876} = \frac{1.36808}{-0.75876} = -1.80305$$

The final result is negative because a positive number (1.36808) is divided by a negative number (-0.75876). The tangents of angles greater than 90 degrees and smaller than 180 degrees are negative. To find an angle whose tangent is negative, find in this case the value nearest to 1.80305 in the columns of tangents in the Handbook tables. It will be seen that the nearest value is 1.8028, which is the tangent of 60° 59′. As the tangent here is negative, angle A, however, is not 60° 59′, but equals 180° − 60° 59′ = 119° 1′. Now angle B is found by the formula

$$B = 180° - (A + C) = 180° - (119° 1' + 20°)$$
$$= 180° - 139° 1' = 40° 59'$$

When Either of Two Triangles Conforms to the Given Dimensions.—When two sides and the angle opposite one of the given sides are known, *if the side opposite the given angle is shorter than the other given side,* two triangles can be drawn as shown by Fig. 5 which have sides of the required length and the required angle opposite one of the sides. The lengths of the two known sides of each triangle are 8 and 9 inches, and the angle opposite the 8-inch side is 49° 27′ in each case; but it will be seen that the angle B of the lower triangle is very much larger than the corresponding angle of the upper triangle, and there is a great difference in the areas. When two sides and one of the opposite angles are given, the problem is capable of two solutions when (and only when) the side opposite the given angle is shorter than the other given side. When the triangle to be calculated is drawn to scale, it is possible to determine from the shape of the triangle which of the two solutions applies.

Example 5:—Find angle B, Fig. 5, from the formula, $\sin B = \dfrac{b \times \sin A}{a}$ where $b = 9$ inches; $A = 49$ degrees, 27 minutes; a is the side opposite angle $A = 8$ inches.

$$\sin B = \frac{9 \times 0.75984}{8} = 0.85482 = \sin 58° 44' 24'' \text{ or } \sin 121° 15'$$

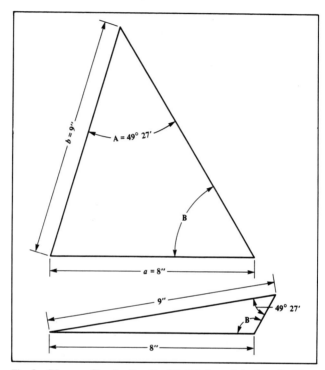

Fig. 5. Diagrams Showing Two Possible Solutions of the Same Problem, which is to Find Angle *B*

36″. The practical requirements of the problem doubtless will indicate which of the two triangles shown in Fig. 5 is the right one.

Example 6:—In Fig. 6, $a = 2$ inches, $b = 3$ inches and $A = 30$ degrees. Find B.

$$\sin B = \frac{b \times \sin A}{a} = \frac{3 \times \sin 30°}{2} = 0.75000$$

We find from the tables that sine 0.75000 is the sine of 48° 35′. From Fig. 6 it is apparent, however, that B is greater than 90 degrees, and as 0.75000 is the sine not only of 48° 35′, but also of 180° − 48° 35′ = 131° 25′, angle B in the case equals 131° 25′.

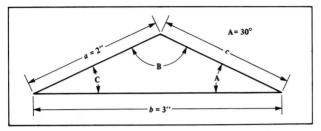

Fig. 6. Another Example which has Two Possible Solutions

This example illustrates how the practical requirements of the problem, indicate which of two angles is correct.

PRACTICE EXERCISES FOR SECTION 9
For answers to all practice exercise problems or questions
see Section 22

1. Three holes in a jig are located as follows:
Hole No. 1 is 3.375 from hole No. 2 and 5.625 from hole No. 3; the distance between No. 2 and No. 3 is 6.250. What three angles between the center lines are thus formed?

2. In Fig. 7 is shown a triangle one side of which is 6.5 feet, and the two angles A and C are 78 and 73 degrees, respectively. Find angle B, sides b and c, and the area.

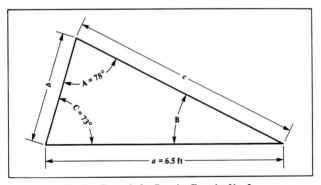

Fig. 7. Example for Practice Exercise No. 2

3. In Fig. 8, side *a* equals 3.2 inches, angle *A*, 118 degrees and angle *B* 40 degrees. Find angle *C*, sides *b* and *c*, and the area.

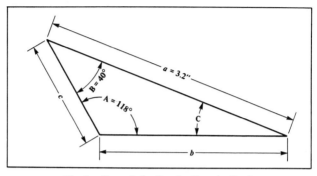

Fig. 8. Example for Practice Exercise No. 3

4. In Fig. 9, side $b = 0.3$ foot, angle $B = 35° 40'$, and angle $C = 24° 10'$, Find angle *A*, sides *a* and *c*, and the area.

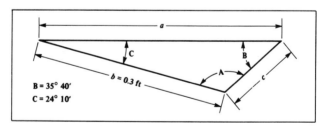

Fig. 9. Example for Practice Exercise No. 4

5. Give two general rules for finding the areas of triangles.

FIGURING TAPERS

The term "taper," as applied in shops and drafting-rooms, means the difference between the large and small dimensions where the increase in size is uniform. Since tapering parts generally are conical, taper means the difference between the large and small diameters. Taper is ordinarily expressed as a certain number of inches per foot; thus ½″ per ft.; ¾″ per ft.; etc. In certain kinds of work, taper is also expressed as a decimal part of an inch per inch, as: 0.050″ per inch. The length of the work is always measured parallel to the center-line (axis) of the work, and never along the tapered surface.

Suppose that the diameter at one end of a tapering part is one inch, and the diameter at the other end, one and one-half inches, and that the length of the part is one foot. This piece, then, tapers one-half inch per foot, because the difference between the diameters at the ends is one-half inch. If the diameters at the ends of a part are ⁷⁄₁₆ inch and ½ inch, and the length is one inch, this piece tapers ¹⁄₁₆ inch per inch. The usual problems met with in regard to figuring tapers may be divided into seven classes. The rule to be used in each case may be found on Handbook page 686.

Example 1:—The diameter at the large end of a part is 2⅝ inches, the diameter at the small end, 2³⁄₁₆ inches, and the length of the work, 7 inches. Find the taper per foot.

Referring to the third rule on Handbook page 686,

$$\text{Taper per foot} = \frac{2\frac{5}{8} - 2\frac{3}{16}}{7} \times 12 = \text{¾ inch}$$

Example 2:—The diameter at the large end of a tapering part is 1⅝ inches, the length is 3½ inches, and the taper per foot is ¾ inch. The problem is to find the diameter at the small end.

Applying the fourth rule on Handbook page 686.

$$\text{Diameter at small end} = 1\frac{5}{8} - \left(\frac{\frac{3}{4}}{12} \times 3\frac{1}{2} \right) = 1\frac{13}{32} \text{ inch}$$

71

Example 3:—What is the length of the taper if the two end diameters are 2.875 inches and 2.542 inches, the taper per foot being one inch?

Applying the sixth rule on Handbook page 686,

Distance between the two diameters = $\dfrac{2.875 - 2.542}{1} \times 12 = 4$ inches nearly

Example 4:—If the length of the taper is 10 inches and the taper per foot is ¾ inch, what is the taper in the given length?

Applying the last rule on Handbook page 686,

$$\text{Taper in given length} = \dfrac{\frac{3}{4}}{12} \times 10 = 0.625 \text{ inch}$$

Example 5:—The small diameter is 1.636 inches, the length of the work is 5 inches, and the taper per foot is ¼ inch; what is the large diameter?

Referring to the fifth rule on Handbook page 686,

$$\text{Diameter at large end} = \left(\dfrac{\frac{1}{4}}{12} \times 5 \right) + 1.636 = 1.740 \text{ inches}$$

Example 6:—Sketch *A*, Fig. 1, shows a part used as a clamp bolt. The diameter, 3¼ inches, is given 3 inches from the large end of the taper. The total length of the taper is 10 inches. The taper is ⅜ inch per foot. Find the diameters at the large and small ends of the taper.

First find the diameter of the large end using the fifth rule on Handbook page 686.

$$\text{Diameter at large end} = \left(\dfrac{\frac{3}{8}}{12} \times 3 \right) + 3\tfrac{1}{4} = 3\tfrac{11}{32} \text{ inches}$$

To find the diameter at the small end, use the fourth rule on the Handbook page mentioned.

$$\text{Diameter at small end} = 3\tfrac{11}{32} - \left(\dfrac{\frac{3}{8}}{12} \times 10 \right) = 3\tfrac{1}{32} \text{ inches}$$

Example 7:—At *B*, Fig. 1, is shown a taper master gage intended for inspecting taper ring gages of various dimensions. The smallest diameter of the smallest ring gage is 1¾ inches, and the largest di-

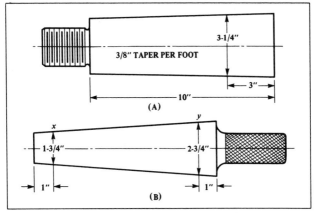

Fig. 1. Illustrations for Examples 6 and 7

ameter of the largest ring gage is 2¾ inches. The taper per foot is 1½ inches. It is required that the master gage extend one inch outside of the ring gages at both the small and the large ends, when these ring gages are tested. How long should the taper on the master gage be?

The sixth rule on Handbook pages 1583 may be applied here.

$$\text{Distance between the two diameters} = \frac{2\frac{3}{4} - 1\frac{3}{4}}{1\frac{1}{2}} \times 12 = 8 \text{ inches}$$

Total length of taper = 8 + 2 = 10 inches

Table for Converting Taper Per Foot to Degrees.—Some types of machines, such as milling machines, are graduated in degrees, making it necessary to convert the taper per foot to the corresponding angle in degrees. This is quickly done by means of the table, Handbook page 687.

Example 8:—If a taper of 1½ inches per foot is to be milled on a piece of work, at what angle must the machine table be set if the taper is measured from the axis of the work?

Referring to the Handbook table, the angle corresponding with a taper of 1½ inches to the foot is 3° 34' 35" as measured from the center line.

Note that the taper per foot varies directly as *the tangent of one-half the included angle*. Two mistakes frequently made in figuring tapers are assuming that the taper per foot varies directly as the included angle or that it varies directly as the tangent of the included angle. In order to verify this point, refer to the table on Handbook page 687 where it will be seen that the included angle for a taper of 4 inches per foot (18° 55′ 29″) is not twice the included angle for a taper of 2 inches per foot (9° 31′ 38′). Neither is the tangent of 18° 55′ 29″ (0.3428587) twice the tangent of 9° 31′ 38″ (0.1678311).

Tapers for Machine Tool Spindles.—The holes in machine tool spindles, for receiving tool shanks, arbors and centers, are tapering to insure a tight grip, accuracy of location, and also to facilitate removal of arbors, cutters, etc. The most common tapers are the Morse, the Brown & Sharpe, and the Jarno. The Morse has been very generally adopted for drilling machine spindles. Most engine lathe spindles also have the Morse taper, but some lathes have the Jarno or a modification of it, and others, a modified Morse taper which is longer than the standard. A standard milling machine spindle was adopted in 1927 by the milling machine manufacturers of the National Machine Tool Builders' Association. A comparatively steep taper of 3½ inches per foot was adopted in connection with this standard spindle to insure instant release of arbors. Prior to the adoption of the standard spindle, the Brown & Sharpe taper was used for practically all milling machines and this is also the taper for dividing-head spindles. There is considerable variation in grinding machine spindles. The Brown & Sharpe taper is the most common, but the Morse and the Jarno have also been used. Tapers of ⅝ inch per foot and ¾ inch per foot have also been used to some extent on miscellaneous classes of machines requiring a taper hole in the spindle.

PRACTICE EXERCISES FOR SECTION 10
For answers to all practice exercise problems or questions
see Section 22

1. What tapers, per foot, are used with the following tapers: (*a*) Morse taper; (*b*) Jarno taper; (*c*) milling machine spindle; (*d*) taper pin?

2. What is the taper per foot on a part if the included angle is 10° 30′; 55° 45′?

3. In setting up a taper gage like that shown on Handbook page 688, what should be the center distance between 1.75" and 2" disks to check either the taper per foot or angle of a No. 4 Morse taper?

4. If it is required to check an angle of 14½°, using two disks in contact, and the smaller disk is 1" diameter, what should be the diameter of the larger disk?

5. What should be the center distance, using disks of 2" and 3" diameter, to check an angle of 18° 30', if the taper is measured from one side?

6. In grinding a reamer shank to fit a standard No. 2 Morse taper gage, it was found that the reamer lacked ⅜ of an inch of going into the gage to the gage mark. How much should be ground off of the diameter?

7. A milling machine arbor has a shank 6½ inches long with a No. 10 B. & S. taper. What is the total taper in this length?

8. A taper bushing for a grinding machine has a small inside diameter of ⅞". It is 3" long with ½" taper per foot. Find the large inside diameter.

9. If a 5-inch sine bar is used for finding the angle of the tapering block A (Fig. 2) and the heights of the sine-bar plugs are as shown, find the corresponding angle a by means of the table beginning on Handbook page 678.

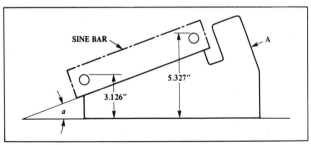

Fig. 2. Finding Angle a by Means of a Sine Bar and Handbook Table

TOLERANCES AND ALLOWANCES FOR MACHINE PARTS

In manufacturing machine parts according to modern methods, certain maximum and minimum dimensions are established, particularly for the more important members of whatever machine or mechanism is to be constructed. These limiting dimensions serve two purposes: they prevent both unnecessary accuracy and excessive inaccuracies. A certain degree of accuracy is essential to the proper functioning of the assembled parts of a mechanism, but it is useless and wasteful to make parts more precise than needed to meet practical requirements; hence, the use of proper limiting dimensions promotes efficiency in manufacturing, and insures standards of accuracy and quality that are consistent with the functions of the different parts of a mechanical device.

Parts made to specified limits usually are considered interchangeable or capable of use without selection but there are several degrees of interchangeability in machinery manufacture. Strictly speaking, interchangeability consists in making the different parts of a mechanism so uniform in size and contour that each part of a certain model will fit any mating part of the same model, regardless of the lot to which it belongs or when it was made. However, as often defined, interchangeability consists in making each part fit any mating part in a certain series; that is, the interchangeability exists only in the same series. Selective assembly is sometimes termed interchangeability, but it involves a selection or sorting of parts as explained later. It will be noted that the strict definition of interchangeability does not imply that the parts must always be assembled without hand work, although that is usually considered desirable. It does mean, however, that when the mating parts are finished, by whatever process, they must assemble and function properly, without fitting individual parts one to the other.

When a machine having interchangeable parts, has been installed possibly at some distant point, a broken part can readily be replaced by a new one sent by the manufacturer, but this feature is secondary as compared with the increased efficiency in manufacturing on an interchangeable basis. In order to make parts interchangeable, it is necessary to use gages and measuring tools, to provide some system of inspection, and to adopt suitable tolerances. Whether absolute interchangeability is practicable or not may depend upon the tolerances adopted, the relation between the different parts, and their form.

Meanings of the Terms "Limit," "Tolerance," and "Allowance."—The terms "limit" and "tolerance" and also "tolerance" and "allowance" are often used interchangeably, but these three terms each has a distinct meaning and refers to different dimensions. As shown by the accompanying diagram, Fig. 1, the *limits* of a hole

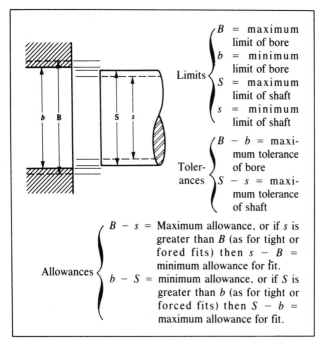

Limits
B = maximum limit of bore
b = minimum limit of bore
S = maximum limit of shaft
s = minimum limit of shaft

Tolerances
$B - b$ = maximum tolerance of bore
$S - s$ = maximum tolerance of shaft

Allowances
$B - s$ = Maximum allowance, or if s is greater than B (as for tight or fored fits) then $s - B$ = minimum allowance for fit.
$b - S$ = minimum allowance, or if S is greater than b (as for tight or forced fits) then $S - b$ = maximum allowance for fit.

Fig. 1.   Diagram Showing Difference Between "Limit," "Tolerance," and "Allowance"

or shaft are its diameters. *Tolerance* is the difference between two *limits* or limiting dimensions of a given part, and the term means that a certain amount of error is tolerated for practical reasons. *Allowance* is the difference between limiting dimensions on mating parts which are to be assembled either loosely or tightly, depending upon the amount allowed for the fit.

Example 1:—Limits and fits for cylindrical parts are given on pages 630 to 657 in the Handbook. This data provides a series of standard types and classes of fits. From the table on page 636, establish limits of size and clearance for a 2-inch diameter hole and shaft for a class RC-1 fit (hole H5, shaft g4).

Max. hole = 2 + 0.0005 = 2.0005; min. hole = 2 − 0 = 2.

Max. shaft = 2 − .0004 = 1.9996; min. shaft = 2 − 0.0007 = 1.9993.

Min. allow. = min. hole − max. shaft = 2 − 1.9996 = 0.0004.

Max. allow. = max. hole − min. shaft = 2.0005 − 1.9993 = 0.0012.

Example 2:— Beginning on Handbook page 1498, there are tables of dimensions for the Standard Unified Screw Thread Series—Class 1A, 2A and 3A and B Fits. Determine the pitch-diameter tolerance of both screw and nut, and also the minimum and maximum allowance between screw and nut at the pitch diameter, assuming that the nominal diameter is 1 inch, the pitch is 8 threads per inch, and the fits are Class 2A and 2B for screw and nut respectively.

The maximum pitch diameter or limit of the screw = 0.9168, and the minimum pitch diameter, 0.9100; hence, the tolerance = 0.9168 − 0.9100 = 0.0068 inch. The nut tolerance = 0.9276 − 0.9100 = 0.0176 inch. The maximum allowance for medium fit = maximum pitch diameter of nut − minimum pitch diameter of screw = 0.9276 − 0.9168 = 0.0108 inch. The minimum allowance = minimum pitch diameter of nut − maximum pitch diameter of screw = 0.9188 − 0.9168 = 0.0020.

Relation of Tolerances to Limiting Dimensions and How Basic Size is Determined.—The absolute limits of the various dimensions and surfaces indicate danger points, inasmuch as parts made beyond these limits are unserviceable. A careful analysis of a mechanism shows that one of these danger points is more sharply defined than the other. For example, a certain stud must always assemble into a certain hole. If the stud is made beyond its maximum limit, it may be too large to assemble. If it is made beyond its minimum limit, it may be too loose or too weak to function. The absolute maximum limit in this case may cover a range of 0.001 inch, whereas the absolute minimum limit may have a range of at least 0.004 inch. In this case the maximum limit is the more sharply defined.

The basic size expressed on the component drawing is that limit which defines the more vital of the two danger points, while the tolerance defines the other. In general, the basic dimension of a male part such as a shaft is the maximum limit which requires a minus tolerance. Similarly, the basic dimension of a female part is the minimum limit requiring a plus tolerance, as shown in Fig. 2. There are, however, dimensions which define neither a male nor a female surface, such for example as dimensions for the location of holes. In a few cases of this kind, a variation in one direction is less dangerous than a variation in the other. Under these conditions, the basic dimension represents the danger point, and the unilateral tolerance permits a variation only in the less dangerous direction. At other times, the conditions are such that any variation from a fixed point in either direction is equally dangerous. In such a case, the basic size represents this fixed point and tolerances on the drawing are bilateral and extend equally in both directions. (See Handbook page 623 for explanation of unilateral and bilateral tolerances.)

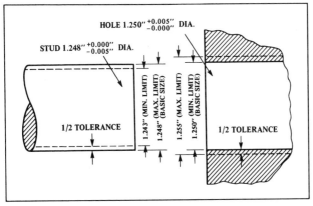

Fig. 2.   Graphic Illustration of the Meaning of the Term Basic Size or Dimension

When Allowance Provides Clearance Between Mating Parts.—
When one part must fit freely into another part like a shaft in its bearing, the allowance between the shaft and bearing represents a clearance space. It is evident that the amount of clearance varies widely for different classes of work. The minimum clearance should be as small as will permit the ready assembly and operation of the parts, while the maximum clearance should be as great as the functioning of the mechanism will allow. The difference between the maximum and minimum clearances defines the extent of the tolerances. In general, the difference between the basic sizes of companion parts equals the minimum clearance (see Fig. 3), and the term "allowance," if not defined as maximum or minimum, is quite commonly applied to the minimum clearance.

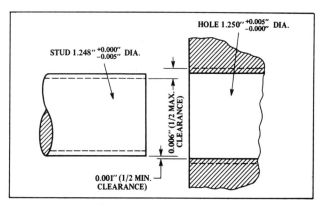

Fig. 3.   Graphic Illustration of the Meaning of the Terms Maximum and Minimum Clearance

When "Interference of Metal" is Result of Allowance.—If a shaft or pin is larger in diameter than the hole into which it is forced, there is, of course, interference between the two parts. The metal surrounding the hole is expanded and compressed as the shaft or other part is forced into place. Engine crankpins, car axles and various other parts are assembled in this way (see paragraph "Allowance for Forced Fits," Handbook page 625). The force and shrink fits in Table 6 (pages 642 and 643) all represent interference of metal.

If interchangeable parts are to be forced together, the minimum interference establishes the danger point. This means that for force fits the basic dimension of the shaft or pin is the minimum limit requiring a plus tolerance, while the basic dimension of the hole is the maximum limit requiring a minus tolerance. (See Fig. 4.)

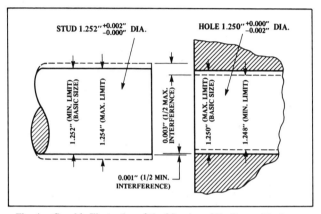

Fig. 4. Graphic Illustration of the Meaning of the Terms Maximum and Minimum Interference

Obtaining Allowance by Selection of Mating Parts.—The term "selective assembly" is applied to a method of manufacturing which is similar in many of its details to interchangeable manufacturing. In selective assembly, the mating parts are sorted according to size, and assembled or interchanged with little or no machining. The chief purpose of manufacturing by selective assembly is the production of large quantities of duplicate parts as economically as possible. As a general rule, the smaller the tolerances, the more exacting and expensive will be the manufacturing processes; but it is possible to use comparatively large tolerances and then reduce them, in effect, by selective assembly, provided the quantity of parts is large enough to make such selective fitting possible. To illustrate the procedure, Fig. 5 shows a plug or stud which has a plus tolerance of 0.001 inch and a hole which also has a plug tolerance of 0.001 inch. Assume that this tolerance of 0.001 inch represents the normal size variation on each part when manufactured efficiently. With this tolerance, a minimum plug in a maximum hole would have a clearance of 0.2510

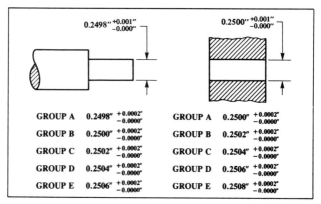

Fig. 5.   Information Placed on Drawings Used in Selective Assembly Manufacturing to Facilitate Grading of Parts

— 0.2498 = 0.0012 inch; and a maximum plug in a minimum hole would have a "metal interference" of 0.2508 — 0.2500 = 0.0008 inch. But suppose the clearance required for these parts must range from zero to 0.0004 inch. This reduction can be obtained by dividing both plugs and holes into five groups. (See Fig. 5.) Any studs in Group A, for example, will assemble in any hole in Group A, but the studs in one group will not assemble properly in the holes in another group. When the largest stud in Group A is assembled in the smallest hole in Group A, the clearance equals zero. When the smallest stud in Group A is assembled in the largest hole in Group A, the clearance equals 0.0004 inch. Thus, in selective assembly manufacturing, there is a double set of limits, the first being the manufacturing limits, and the second the assembling limits. In many cases two separate drawings are made of a part which is to be graded before assembly. One shows the manufacturing tolerances only, so as not to confuse the operator, while the other gives the proper grading information.

Example 3:—Data for force and shrink fits are given in the table on page 642 in the Handbook. Establish the limits of size and interference of the hole and shaft for a Class FN-1 fit of 2-inch diameter.

Max. hole = 2 + 0.0007 = 2.0007; min. shaft = 2 − 0 = 2.

Max. shaft = 2 + 0.0018 = 2.0018; min. shaft = 2 + 0.0013 = 2.0013.

In the second column of the table the minimum and maximum interference are given as 0.0006 and 0.0018 inch, respectively, for a FN-1 fit of 2-inch diameter. For a "selected" fit, shafts are selected which are 0.0012 inch larger than the mating holes; that is, for any mating pair the shaft is larger than the hole by an amount midway between the minimum (0.0006 inch) and maximum (0.0018 inch) interference.

Dimensioning Drawings to Ensure Obtaining Required Tolerances.—In dimensioning the drawings of parts requiring tolerances, there are certain fundamental rules that should be applied

Rule 1.   In interchangeable manufacturing there is only one dimension (or group of dimensions) in the same straight line which can be controlled within fixed tolerances. This is the distance between the cutting surface of the tool and the locating or registering surface of the part being machined. Therefore, it is incorrect to locate any point or surface with tolerances from more than one point in the same straight line.

Rule 2.   Dimensions should be given between those points which it is essential to hold in a specific relation to each other. The majority of dimensions, however, are relatively unimportant in this respect. It is good practice to establish common location points in each plane and to give, as far as possible, all such dimensions from these points.

Rule 3.   The basic dimensions given on component drawings for interchangeable parts should be, except for force fits and other unusual conditions, the "maximum metal" sizes (maximum shaft or plug and minimum hole.) The direct comparison of the basic sizes should check the danger zone, which is the minimum clearance condition in the majority of cases. It is evident that these sizes are the most important ones, as they control the interchangeability, and they should be the first determined. Once established, they should remain fixed if the mechanism functions properly and the design is unchanged. The direction of the tolerances, then, would be such as to recede from the danger zone. In the majority of cases, this means that the direction of the tolerances is such as will increase the clearance. For force fits, the basic dimensions determine the minimum interference, while the tolerances limit the maximum interference.

Rule 4.   Dimensions must not be duplicated between the same points. The duplication of dimensions causes much needless trouble, due to changes being made in one place and not in the others. It causes less trouble to search a drawing to find a dimension than it does to have them duplicated and more readily found but inconsistent.

Rule 5. As far as possible, the dimensions on companion parts should be given from the same relative locations. Such a procedure assists in detecting interferences and other improper conditions.

In attempting to work in accordance with general laws or principles, one other elementary rule should always be kept in mind. Special cases require special consideration. The following detailed examples are given to illustrate the application of these five laws and to indicate results of their violation.

Violations of Rules for Dimensioning.—Fig. 6 shows a very common method of dimensioning a part such as the stud shown, but one that is bad practice. it violates the first and second rules. As the dimensions given for the diameters are correct, they are eliminated from the discussion. The dimensions given for the various lengths are wrong: First, because they give no indication as to the essential lengths; second, because of several possible sequences of operations, some of which would not maintain the specified conditions.

Fig. 7 shows one possible sequence of operations indicated alphabetically. If we first finish the dimension *a* and then finish *b*, the dimension *c* will be within the specified limits. In this case, however, the dimension *c* is superfluous. Fig. 8 gives another possible sequence of operations. If we first establish *a*, and then *b*, the dimension *c* may vary 0.030 instead of 0.010 inch as is specified in Fig. 6. Fig. 9 gives a third possible sequence of operations. If we first finish the

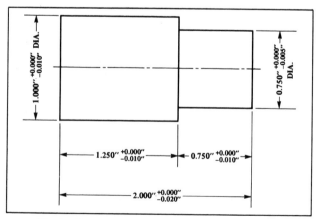

Fig. 6. Common but Incorrect Method of Dimensioning

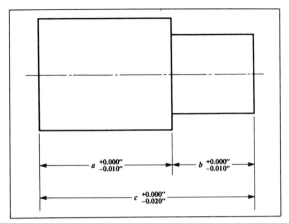

Fig. 7. One Interpretation of Dimensioning in Fig. 6

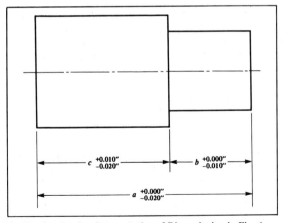

Fig. 8. Another Interpretation of Dimensioning in Fig. 6

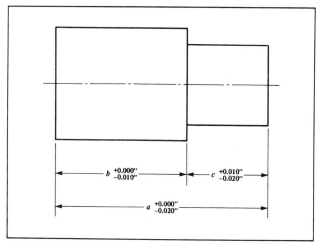

Fig. 9. A Third Interpretation of Dimensioning in Fig. 6

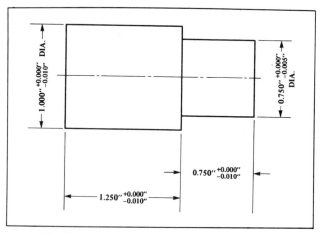

Fig. 10. Correct Dimensioning if Length of Body and Length of Stem are Most Important

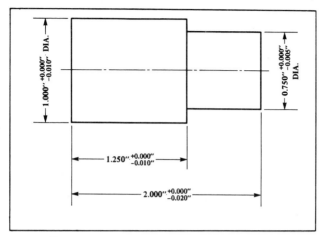

Fig. 11. Correct Dimensioning if Length of Body and Over-all Length are Most Important

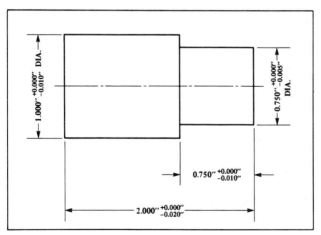

Fig. 12. Correct Dimensioning if Over-all Length and Length of Stem are Most Important

over-all length a, and then the length of the body b, the stem c may vary 0.030 inch instead of 0.010 inch as specified in Fig. 6.

If three different plants were manufacturing this part, each one using a different sequence of operations, it is evident from the foregoing that a different product would be received from each plant. The example given is the simplest one possible. As the parts become more complex, and the number of dimensions increase, the number of different combinations possible and the extent of the variations in size that will develop also increase.

Fig. 10 shows the correct way to dimension this part if the length of the body and the length of the stem are the essential dimensions. Fig. 11 is the correct way if the length of the body and the length over all are the most important. Fig. 12 is correct if the length of the stem and the length over all are the most important. If the part is dimensioned in accordance with either Fig. 10, Fig. 11, or Fig. 12, the product from any number of factories should be alike.

PRACTICE EXERCISES FOR SECTION 11
For answers to all practice exercise problems or questions
see Section 22

1. What factors influence the allowance for a forced fit?

2. What is the general practice in applying tolerances to center distances between holes?

3. A 2-inch shaft is to have a tolerance of 0.003 inch on the diameter. Show, by examples, three ways of expressing the shaft dimensions.

4. In what respect does a bilateral tolerance differ from a unilateral tolerance; show, by example?

5. What are the standard thickness tolerances for cold-rolled sheets?

6. What are the tolerances for hexagonal hot-rolled carbon steel bars?

7. What is the relationship between gagemaker's tolerance and workpiece tolerance?

8. Name the different classes of fits for screw threads included in the American Standard.

9. How does the Unified Standard for Screw Threads differ from the former American Standard with regard to clearance between mating parts? With regard to working tolerance?

10. Under what conditions is one limiting dimension or "limit" also a basic dimension?

11. What do the letter symbols RC, LC, LT, LN and FN signify with regard to American Standard fits?

12. According to the table at the bottom of page 632, broaching will produce work within tolerance grades 5 through 8. What does this mean in terms of thousandths of an inch, considering a 1-inch diameter broached hole?

13. Does surface roughness affect the ability to work within the tolerance grades specified in question 12?

SECTION 12

USING STANDARDS DATA AND INFORMATION

Standards are needed in metalworking manufacturing to establish dimensional and physical property limits for parts that are to be interchangeable. Standards make it possible for parts such as nuts, screws, bolts, splines, gears, etc., to be manufactured at different times and places with the assurance that they will meet assembly requirements. Standards are also needed for tools such as twist drills, reamers, milling cutters, etc., so that only a given number of sizes need be made available to cover a given range and to ensure adequate performance. Also, performance standards are often established to make sure that machines and equipment will satisfy their application requirements.

A standard may be established by a company on a limited basis for its own use. An industry may find that a standard is needed, and its member companies working through their trade association come to an agreement as to what requirements should be included. Sometimes industry standards sponsored by a trade association or an engineering society become accepted by a wide range of consumers, manufacturers, and government agencies as national standards and are made available through a national agency such as the American National Standards Institute (ANSI). More and more countries are coming to find that standards should be universal and are working to this end through the International Standards Organization (ISO).

In the United States and some other English-speaking countries there are two systems of measurement in use: the inch system and the metric system. As a result, standards for, say, bolts, nuts, and screws have been developed for both inch and metric dimensions as will be found in MACHINERY'S HANDBOOK. However, an increasing number of multinational corporations and their local suppliers are finding it prohibitively expensive to operate with two systems of measurements and standards. Thus, in order to use available expertise in one plant location, a machine may be designed in an "inch" nation only to be produced later in a "metric" country or vice versa. This situation generates additional costs in the conversion of drawings, substitution of equivalent standard steel sizes and fasteners, the conversion of testing and material specifications, etc. Because of these problems, more and more standards are being developed in the United States and throughout the world that are based, wherever practicable, upon ISO standards.

In the Handbook, the user will find that a large number of both inch and metric standards data and information are provided. It should be noted that at the head of each table of standards data the source is given in parentheses, such as (ANSI B18.3-1982). The ANSI indicates the American National Standards Institute; the B18.3 is the identifying number of the standard; and the 1982 is the date the standard was published and became effective.

Generally, new products are produced to the metric standards; older products and replacement parts for them may require reference to older inch standards; and some products such as inch-unit pipe threads are considered as standard for the near future because of widespread use throughout the world.

Important Objectives of Standardization*.—The purpose of standardization is to manufacture goods for less direct and indirect costs and to provide finished products that meet the demands of the marketplace. A more detailed description of the objectives could be as follows:

Lower the production costs when the aim is to:

1. Facilitate and systematize the work of skilled designers;

2. Ensure optimum selection of materials, components, and semifinished products;

3. Reduce stocks of materials, semifinished products, and finished products;

4. Minimize the number of different products sold; and

5. Facilitate and reduce the cost of procurement of purchased goods.

**World Metric Standards for Engineering;* Knut O. Kverneland; Industrial Press, New York.

Meet the demands of the marketplace, when the objective is to:

1. Conform to regulations imposed by government and trade organizations;

2. Stay within safety regulations set forth by governments; and

3. Facilitate interchangeability requirements with existing products.

Standardization Technique.—Two basic principles for the preparation of a standard are commonly used; these are:

1. Analytical standardization—Standard developed from scratch.

2. Conservative standardization—Standard based, so far as is possible, on existing practice.

In practice, it appears that a standard cannot be prepared completely by one or the other of the two methods, but emerges from a compromise between the two. The goal of the standardization technique then should be to utilize the basic material and the rules and the aids available in such a way that a valid and practical compromise solution is reached.

The basic material could consist of such items as: former company standards; vendor catalog data; national and international standards; requirements of the company's customers; and competitor's material. Increasingly important are the national and international standards in existence on the subject; they should always play an important part in any conservative standardization work. For example, it would be foolish to create a new metric standard without first considering some existing European metric standards.

Standards Information in the Handbook.—Among the many kinds of material and data to be found in the Handbook the user will note that extensive coverage is given to standards of several types: American National Standards; British Standards; ISO Standards; trade association standards; and, in certain instances, company product standards. Both inch and metric system standards are given wherever appropriate. In some cases inch dimension standards are provided only for use during transition to metric standards or to provide information for the manufacture of replacement parts.

In selecting standards to be presented in the Handbook, the editors have chosen those standards most appropriate to the needs of Handbook users. Tables of data, formulas, and illustrations have been arranged to be usable directly and quickly. Where combinations of tables and formulas are given, the formulas have been arranged in the best sequence for computation with the aid of a hand calculator. As an example of this type of presentation, the section on fine-pitch worms and wormgears, pages 1876–1885, begins with text material on page 1876 that provides the basis for the information presented in the standard; the illustration and formulas on page 1877 show how the proportions of fine-pitch worms and wormgears are determined; the tables on pages 1879 and 1880 provide key data for a standard series of worms; and the worked-out examples beginning on page 1878 give a step-by-step procedure for selecting and dimensioning the worm and wormgear. Finally, on page 1883 there is shown a method for taking into account differences in the profile of the worm caused by the use of large and small diameter cutting tools.

"Soft" Conversion of Inch to Metric Dimensions.—The dimensions of certain products, when specified in inches, may be converted to metric dimensions, or vice versa, by multiplying by the appropriate conversion factor so that the parts can be fabricated either to inch or to the equivalent metric dimensions and still be fully interchangeable. Such a conversion is called a "soft" conversion. An example of a "soft" conversion is available on page 2163 on which are given the inch dimensions of standard lockwashers for ball bearings. The footnote to the table indicates that multiplication of the tabulated inch dimensions by 25.4 and rounding the results to two decimal places will provide the equivalent metric dimensions.

"Hard" Metric or Inch Standard Systems.—In a "hard" system, those dimensions in the system that have been standardized cannot be converted to another dimensional system that has been standardized independent of the first system. As stated in the footnote on page 2039 of the Handbook, "In a 'hard' system the tools of production, such as hobs, do not bear a usable relationship to the tools in another system; i.e., a 10 diametral pitch hob calculates to be equal to a 2.54 module hob in the metric module system, a hob that does not exist in the metric standard."

Interchangeability of Parts Made to Revised Standards.—In cases where a standard has been revised there may still remain some degree of interchangeability between older parts and those made to the new standard. As an example, on pages 2028 and 2029 of the Handbook there are two tables showing which of the internal and external involute splines made to older standards will mate with those made to newer standards.

PRACTICE EXERCISES FOR SECTION 12

For answers to all practice exercise problems or questions
see Section 22

1. What is the tolerance on thickness for an aluminum sheet 10 inches wide and 0.012 inch thick?

2. What is the breaking strength of a 6 × 7 fiber-core wire rope ¼ inch in diameter if the rope material is mild plow steel?

3. What factor of safety should be applied to the rope in problem 2?

4. How many carbon steel balls of ¼ inch diameter would weigh 1 lb.?

5. For a 1-inch diameter shaft what size square key is appropriate?

6. Find the hole size needed for a ⁵⁄₃₂ inch standard cotter pin.

7. Find the limits of size for a 0.1250-inch diameter hardened and ground dowel pin.

8. For a 3AM1-17 retaining ring (snap ring) what is the maximum allowable speed of rotation?

9. Find the hole size required for a type AB steel thread-forming screw of number 6 size in 0.105-inch-thick stainless steel.

SECTION 13

STANDARD SCREW AND PIPE THREADS

HANDBOOK Pages 1474–1591 and 1611–1620

Different screw-thread forms and standards have been originated and adopted at various times, either because they were considered superior to other forms or because of the special requirements of screws used on a certain class of work.

A standard thread conforms to an adopted standard in regard to the form or contour of the thread itself, and as to the pitches or numbers of threads per inch for different screw diameters. A screw thread having either a modified form or a pitch which is either greater or less for a given screw diameter, than the adopted standard, is special.

The United States Standard formerly used in the United States was replaced by an American Standard having the same thread form as the former standard and a more extensive series of pitches, as well as tolerances and allowances for different classes of fits. This American Standard was revised in 1949 to include a Unified Thread Series which was established to obtain screw-thread interchangeability among the United Kingdom, Canada and the United States.

The Standard was revised again in 1959. The Unified threads are now the standard for use in the United States and the former American Standard threads are now used only in certain applications where the changeover in tools, gages, and manufacturing has not been completed. The differences between Unified and the former National Standard threads are explained on pages 1474 and 1479 in the Handbook.

As may be seen in the table on Handbook page 1483, the Unified Series of screw threads consists of three standard series having graded pitches (UNC, UNF, and UNEF) and eight standard series of uniform (constant) pitch. In addition to these standard series, there are places in the table beginning on page 1498 where special threads (UNS) are listed. These UNS threads are for use only if standard series threads do not meet requirements.

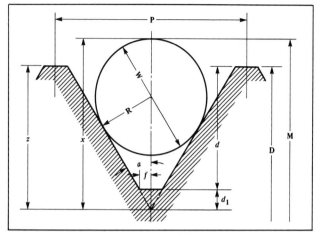

Fig. 1. Diagram Illustrating the Derivation of Formulas for Three-wire Measurements of Screw Thread Pitch Diameters

Example 1:—The table on Handbook page 1484 shows that the pitch diameter of a 2-inch screw thread is 1.8557 inches. What is meant by the term "pitch diameter" as applied to a screw thread and how is it determined?

According to a definition of "pitch diameter" given in connection with American Standard screw threads, the pitch diameter of a straight (non-tapering) screw thread is the diameter of an imaginary cylinder the surface of which would pass through the threads at such points as to make equal the width of the threads and the width of the spaces cut by the surface of the cylinder.

The basic pitch diameter equals the basic major (outside) diameter minus two times the addendum of the external thread (page 1482). In this case, the basic pitch diameter is $2.0000 - 2 \times 0.07217 = 1.8557$ inches.

Example 2:—The tensile strength of a bolt, 3½ inches in diameter at a stress of 6000 pounds per square inch may be calculated by means of the formula on Handbook page 1278. This formula uses the largest diameter of the bolt, avoiding the need to take account of the reduced diameter at the thread root, and gives a tensile strength of 35,175 pounds for the conditions noted.

If the second formula on page 1278, based on the area of the smallest diameter, is used for the same bolt and stress, and the diameter of the thread root is taken as 3.1 inches, then the tensile strength is calculated as 40,636 pounds. The difference in these formulas is that the first uses a slightly greater factor of safety than the second, taking account of possible variations in thread depth.

Example 3:—On Handbook page 1628 formulas are given for checking the pitch diameter of screw threads by the three-wire method (when effect of lead angle is ignored). Show how these formulas have been derived, using the one for the American National Standard Unified thread as an example.

It is evident from the diagram, Fig. 1, that

$$M = D - 2z + 2x \tag{1}$$

$$x = R + \frac{R}{\sin a} \text{ and } 2x = 2R + \frac{2R}{0.5}; \text{ hence}$$

$$2x = \frac{(2 \times 0.5 + 2)R}{0.5} = \frac{3R}{0.5} = 6R = 3W$$

$$z = d + d_1 = 0.6495P + f \times \cot a$$

$$f = 0.0625P; \text{ therefore}$$

$$z = 0.6495P + 0.10825P = 0.75775P$$

If in formula (1) we substitute the value of $2z$ or $2 \times 0.75775P$ and the value of $2x$, we have

$$M = D - 1.5155 \times P + 3W \tag{2}$$

This formula (2) is the one found in previous editions of the Handbook. In the 22nd edition of the Handbook use of the outside diameter D in formula (2) above was eliminated to provide a formula in terms of the pitch diameter E. Such a formula is useful for finding the wire measurement corresponding to the actual pitch diameter whether it be correct, undersize, or oversize.

According to the last paragraph of the previous example, $E = D - 2 \times$ thread addendum. On Handbook page 1482 the formula for thread addendum given at the top of the last column is $0.32476P$. Therefore, $E = D - 2 \times 0.32476P$, or, transposing this formula, D

$= E + 2 \times 0.32476P = E + 0.64952P$. Substituting this value of D into formula (2) gives: $M = E + 0.64952P - 1.5155P + 3W = E - 0.8660P + 3W$ which is the Handbook formula.

Example 4:—On Handbook page 1638 is given a formula for checking the angle of a screw thread by a three-wire method. How is this formula derived? Referring to the diagram, Fig. 2

$$\sin a = \frac{W}{S} \tag{1}$$

If D = diameter of larger wires and d = diameter of smaller wires

$$W = \frac{D - d}{2}$$

If B = difference in measurement over wires, then the difference S between the centers of the wires is:

$$S = \frac{B - (D - d)}{2}$$

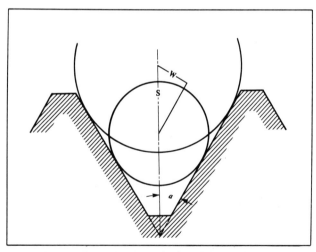

Fig. 2. Diagram Illustrating the Derivation of Formula for Checking the Thread Angle by the Three-wire System

By inserting these expressions for W and S in formula (1) and cancelling, the formula given in the Handbook is obtained if A is substituted for $D - d$.

$$\sin a = \frac{A}{B - A} \qquad (2)$$

Example 5:—A vernier gear-tooth caliper (like the one shown on Handbook page 1791) is to be used for checking the width of an Acme screw by measuring squarely across or perpendicular to the thread. Since standard screw thread dimensions are in the plane of the axis, how is the width square or normal to the sides of the thread determined? Assume that the width is to be measured at the pitch line and that the number of threads per inch is two.

The table on Handbook page 1560 shows that for two threads per inch, the depth is 0.260 inch; hence, if the measurement is to be at the pitch line, the vertical scale of the caliper is set to (0.260 − 0.010) ÷ 2 = 0.125 inch. The pitch equals

$$\frac{1}{\text{No. of Threads per Inch}} = \tfrac{1}{2} \text{ inch}$$

The width A, Fig. 3, in the plane of the axis equals $\frac{1}{2}$ of the pitch, or $\frac{1}{4}$ inch. The width B perpendicular to the sides of the thread = width in axial plane × cosine helix angle. (The helix angle which equals angle a, is based upon the pitch diameter and is measured from a plane perpendicular to the axis of the screw thread.) The

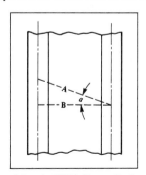

Fig. 3. Determining the Width Perpendicular to the Sides of a Thread at the Pitch Line

width A in the plane of the axis represents the hypotenuse of a right triangle, and the required width B equals the side adjacent; hence width $B = A \times$ cosine of helix angle. The angle of the thread itself (29° for an Acme Thread) does not affect the solution.

Width of Flat End of Unified Screw Thread and American Standard Acme Screw Thread Tools.—The widths of flat or end of the threading tool for either of these threads may be measured by using a micrometer as illustrated at A, Fig. 4. In measuring the thread tool, a scale is held against the spindle and anvil of the micrometer and the end of the tool is placed against this scale. The micrometer is then adjusted to the position shown and 0.2887 inch subtracted from the reading for an American Standard screw thread tool. For the American Standard Acme threads, 0.1293 inch is subtracted from the micrometer reading to obtain the width of the tool point. The constants (0.2887 and 0.1293) which are subtracted from the micrometer reading are only correct when the micrometer spindle has the usual diameter of 0.25 inch.

An ordinary gear-tooth vernier caliper may also be used for testing the width of a thread tool point, as illustrated at B. If the measurement is made at a vertical distance x of ¼ inch from the points of the caliper jaws, the constants previously given for American Standard and American Standard Acme threads should be subtracted from the caliper reading to obtain the actual width of the cutting end of the tool.

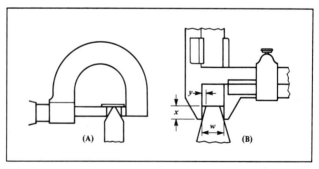

Fig. 4.   Measuring Width of Flat on Threading Tool (A) with a Micrometer; (B) with a Gear Tooth Vernier

Example 6:—Explain how the constants 0.2887 and 0.1293 referred to in a preceding paragraph are derived and deduce a general rule applicable regardless of the micrometer spindle diameter or vertical dimension x, Fig. 4.

The dimension x (which also is equivalent to the micrometer spindle diameter) represents one side of a right triangle (the side adjacent), having an angle of $29 \div 2 = 14$ degrees and 30 minutes, in the case of an Acme thread. The side opposite, or $y = $ side adjacent $\times$ tangent = dimension $x \times$ tan $14° 30'$.

If x equals 0.25 inch, then side opposite or $y = 0.25 \times 0.25862 = 0.06465$; hence, the caliper reading minus $2 \times 0.06465 = $ width of the flat end ($2 \times 0.06465 = 0.1293 = $ constant).

The same result would be obtained by multiplying 0.25862 by $2x$; hence, the following rule: To determine the width of the end of the threading tool, by the general method illustrated in Fig. 4, multiply twice the dimension x (or spindle diameter in the case of the micrometer) by the tangent of one-half the thread tool angle, and subtract this product from the width w to obtain the width at the end of the tool.

Example 7:—A gear tooth vernier caliper is to be used for measuring the width of the flat of an American Standard external screw thread tool. The vertical scale is set to ⅛ inch (corresponding to the dimension x, Fig. 4). How much is subtracted from the reading on the horizontal scale, to obtain the width of the flat end of the tool?

$$\frac{1}{8} \times 2 \times \tan 30° = \frac{1}{4} \times 0.57735 = 0.1443 \text{ inch}$$

Hence, the width of the flat equals w, Fig. 4, minus 0.1443. This width should be equal to one-eighth of the pitch of the thread to be cut, since this is the width of flat at the minimum minor diameter of American Standard external screw threads.

PRACTICE EXERCISES FOR SECTION 13
For answers to all practice exercise problems or questions
see Section 22

1. What form of screw thread is most commonly used (*a*) in the United States? (*b*) in England?

2. What is the meaning of the abbreviations 3″—4NC–2?

3. What are the advantages of an Acme thread as compared with a square thread?

4. For what reason would a Stub Acme thread be preferred in some applications?

5. Find the pitch diameters of the following screw threads of American Standard Unified form: ¼ − 28, (meaning ¼ inch diameter and 28 threads per inch); ¾ − 10?

6. How much taper has a standard pipe thread?

7. Under what conditions are straight, or non-tapering pipe threads used?

8. In cutting a taper thread, what is the proper position for the lathe tool?

9. If a lathe is used for cutting a British Standard pipe thread, in what position is the tool set?

10. A thread tool is to be ground for cutting an Acme thread having 4 threads per inch; what is the correct width of the tool at the end?

11. What are the common shop and tool-room methods of checking the pitch diameters of American Standard screw threads requiring accuracy?

12. In using the formula, Handbook page 1628, for measuring an American Standard screw thread by the three-wire method, why should the constant 0.86603 be multiplied by the pitch before subtracting from measurement M, even if not enclosed by parentheses?

13. What is the difference between the pitch and the lead (a) of a double thread? (b) of a triple thread?

14. In using a lathe to cut American Standard Unified threads, what should be the truncations of the tool points and the thread depths for the following pitches: 0.1, 0.125, 0.2 and 0.25 inch?

15. In using the three-wire method of measuring a screw thread, what is the micrometer reading for a ¾ − 12 special thread of American Standard form, if the wires have a diameter of 0.070 inch?

16. Are most nuts made to the United States Standard dimensions?

17. Is there, at the present time, a Manufacturers Standard for bolts and nuts?

18. The American Standard for machine screws includes a coarse-thread series and a fine-thread series as shown by the tables on Handbook pages 1484 and 1485? Which series is commonly used?

19. How is the length (a) of a flat-head or countersunk type of machine screw measured? (b) of a fillister head machine screw?

20. What size tap drill should be used for an American Standard machine screw of No. 10 size, 24 threads per inch?

21. What is the diameter of a No. 10 drill?

22. Is a No. 6 drill larger than a No. 16?

23. What is the relation between the letter size drills and the numbered sizes?

24. Why is it common practice to use tap drills that leave about ¾ of the full thread depth after tapping, as shown by the tables on pages 1660 and 1661?

25. What form of screw thread is used on (a) machine screws? (b) cap screws?

26. What standard governs the pitches of cap-screw threads?

27. What form of thread is used on the National Standard fire hose couplings? How many standard diameters are there?

28. In what way do hand taps differ from machine screw taps?

29. What are tapper taps?

30. The diameter of a ¾ − 10 American Standard thread is to be checked by the three-wire method. What is the largest size wire that can be used?

31. Why is the advance of some threading dies positively controlled by a lead-screw instead of relying upon the die to lead itself?

32. What is the included angle of the heads of American Standard (a) flat-head machine screws? (b) flat-head cap-screws? (c) flat-head wood screws?

SECTION 14

PROBLEMS IN MECHANICS

HANDBOOK Pages 129–203

In the design of machines or other mechanical devices, it is often necessary to deal with the actions of forces and their effects. For example, the problem may be to determine what force is equivalent to two or more forces acting in the same plane but in different directions. Another type of problem is to determine the change in the magnitude of a force resulting from the application of mechanical appliances such as levers, pulleys and screws used either separately or in combination. It may also be necessary to determine the magnitude of a force in order to proportion machine parts to resist the force safely; or, possibly, to ascertain if the force is great enough to perform a given amount of work. Determining the energy stored in a moving body or its capacity to perform work, and the power developed by mechanical apparatus, or the rate at which work is performed, are additional examples of problems frequently encountered in originating or developing mechanical appliances. That section in MACHINERY'S HANDBOOK on Mechanics, beginning on page 129, deals with fundamental principles and formulas applicable to a large variety of problems of the general classes referred to.

The Moment of a Force.—The tendency of a force acting upon a body is, in general, to produce either a motion of translation (that is, to cause every part of the body to move in a straight line) or to produce a motion of rotation. A *moment,* in mechanics, is the measure of the turning effect of a force which tends to produce rotation. For example, suppose a force acts upon a body which is supported by a pivot. Unless the line of action of the force happens to pass through the pivot, the body will tend to rotate. Its tendency to rotate, moreover, will depend upon two things: (1) the magnitude of the force acting, and (2) the distance of the force from the pivot, *measuring along a line at right angles to the line of action of the force.* (See the diagram at the bottom of Handbook page 137, and the accompanying text.)

104

Example 1:—A force F of 300 pounds is applied to a crank disk A (Fig. 1) and in the direction of the arrow. If the radius $R = 5$ inches, what is the turning moment? Also determine how much the turning moment is reduced when the crankpin is in the position shown by the dotted lines, assuming that the force is along line f and that $r = 2\frac{1}{2}$ inches.

When the crankpin is in the position shown by the full lines, the maximum turning moment is obtained and it equals $F \times R = 300 \times 5 = 1500$ inch-pounds or pound-inches. When the crankpin is in the position shown by the dotted lines, the turning moment is reduced one-half and equals $f \times r = 300 \times 2\frac{1}{2} = 750$ inch-pounds.

Note: Torque or turning moment is sometimes expressed as pound-feet or pound-inches, instead of using the term foot-pounds or inch-pounds. Since *foot-pound* is the unit of work and is used in horsepower calculations, it is considered preferable to reverse it and use the term *pound-foot* to indicate torque or turning moment. The reversal of the term foot-pound serves to distinguish readily between the two units of measurment—the unit of work and the unit of turning moment. The latter ordinarily is expressed as inch-pounds or pound-inches instead of foot-pounds or pound-feet, because the dimensions of shafts and other machine parts ordinarily are given in inches; hence the reversal of the term *inch-pound* is not so important as in the case of foot-pound, because inch-pound is not used as a unit of work.

Example 2:—Assume that the force F (diagram B, Fig. 1) is applied to the crank through a rod connecting with a crosshead which slides

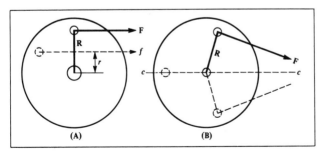

Fig. 1.   Diagram showing how the Turning Moment of a Crank Disk varies from Zero to Maximum

along center-line c-c. If the crank radius $R = 5$ inches, what will be the maximum and minimum turning moments?

The maximum turning moment occurs when the radial line R is perpendicular to the force line F and equals in inch-pounds, $F \times 5$ in this example. When the radial line R is in line with the center-line c-c, the turning moment is o, since $F \times 0 = 0$. This is the "dead-center" position for steam engines and explains why the crankpins on each side of a locomotive are located 90 degrees apart, or, in such a position that the maximum turning moment, approximately, occurs when the turning moment is zero on the opposite side. With this arrangement, it is always possible to start the locomotive since only one side at a time can be in the dead-center position.

The Principle of Moments in Mechanics.—When two or more forces act upon a rigid body and tend to turn it about an axis, then, for equilibrium to exist, the sum of the moments of the forces which tend to turn the body in one direction must be equal to the sum of the moments of those which tend to turn it in the opposite direction about the same axis.

Example 3:—In Fig. 2, a lever 30 inches long is pivoted at the fulcrum F. At the right, and 10 inches from F is a weight, B, of 12 pounds tending to turn the bar in a right-hand direction about its fulcrum F. At the left end, 12 inches from F, the weight of A of 4 pounds tends to turn the bar in a left-hand direction, while weight C, at the other end, 18 inches from F, has a like effect, through the

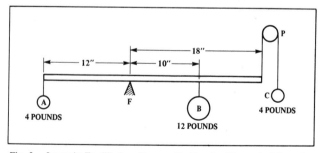

Fig. 2. Lever in Equilibrium Because the Turning Moment of Opposing Forces are Equal

use of the string and pulley P. Taking moments about F, which is the center of rotation, we have:

Moment of $B = 10 \times 12 = 120$ inch-pounds

Opposed to this are the moments of A and C:

Moment of $A = 4 \times 12 = 48$ inch-pounds
Moment of $C = 4 \times 18 = 72$ inch-pounds

Sum of negative moments $= 120$ inch-pounds

Hence, the opposing moments are equal, and, if we suppose, for simplicity, that the lever is weightless, it will balance or be in equilibrium. Should weight A be increased, the negative moments would be greater, and the lever would turn to the left, while if B should be increased or its distance from F be made greater, the lever would turn to the right. (See fourth diagram on Handbook page 137 and accompanying text.)

Example 4:—Another application of the principle of moments is given in Fig. 3. A beam of uniform cross section, weighing 200 pounds, rests upon two supports, R and S, which are 12 feet apart. The weight of the beam is considered to be concentrated at its center of gravity G, at a distance of 6 feet from each support. A weight of 50 pounds is placed upon the beam at a distance of 9 feet from the right-hand support S. Required, the portion of the total weight borne by each support.

Before proceeding, it should be explained that the two supports react or push upward, with a force equal to the downward pressure

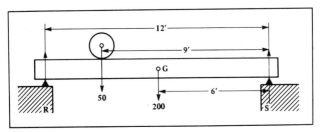

Fig. 3. The Weight on Each Support is Required

of the beam. To make this clear, suppose two men to take hold of the beam, one at each end, and that the supports be withdrawn. Then, in order to hold the beam in position, the two men must together lift or pull upward an amount equal to the weight of the beam and its load, or 250 pounds. Placing the supports in position again, and resting the beam upon them, does not change the conditions. The weight of the beam acts downward, and the supports react by an equal amount.

Now, to solve the problem, assume the beam to be pivoted at one support, say at S. The forces or weights of 50 pounds and 200 pounds tend to rotate the beam in a left-hand direction about this point, while the reaction of R in an upward direction tends to give it a right-hand rotation. As the beam is balanced and has no tendency to rotate, it is in equilibrium, and the opposing moments of these forces must balance; hence, taking moments,

$$9 \times 50 = 450 \text{ pound-feet}$$
$$6 \times 200 = 1200 \text{ pound-feet}$$

Sum of negative moments $= 1650$ pound-feet

Letting R represent the reaction of support,

Moment of $R = R \times 12 =$ pound-feet

By the principle of moments, $R \times 12 = 1650$. That is, if R, the quantity which we wish to obtain, be multiplied by 12, the result will be 1650; hence, to obtain R, divide 1650 by 12, whence $R = 137.5$ pounds, which is also the weight of that end of the beam. As the total load is 250 pounds, the weight at the other end must be $250 - 137.5 = 112.5$ pounds.

The Principle of Work in Mechanics.—There is another principle of more importance than the principle of moments, even in the study of machine elements. It is called the principle of work. According to this principle (neglecting frictional or other losses) the applied force, multiplied by the distance through which it moves, equals the resistance overcome, multiplied by the distance through which it is overcome. The principle of work may also be stated as follows:

Work put in $=$ lost work $+$ work done by machine

This principle holds absolutely in every case. It applies equally to a simple lever, the most complex mechanism, or to a so-called "perpetual motion" machine. No machine can be made to perform work unless a somewhat greater amount—enough to make up for the losses—be applied by some external agent. In the "perpetual motion" machine no such outside force is supposed to be applied, hence such a machine is impossible, and against all the laws of mechanics.

Example 5:—Assume that a rope exerts a pull F of 500 pounds (upper diagram, Handbook page 149) and that the pulley radius $R = 10$ inches and the drum radius $r = 5$ inches. How much weight W can be lifted (ignoring frictional losses) and upon what mechanical principle is the solution based?

According to one of the formulas accompanying the diagram at the top of Handbook page 149,

$$W = \frac{F \times R}{r} = \frac{500 \times 10}{5} = 1000 \text{ pounds}$$

This formula (and the others for finding the values of F, R, etc.) agrees with the principle of moments, and also with the principle of work. The principle of moments will be applied first.

The moment of the force F about the center of the pulley, which corresponds to the fulcrum of a lever, is F multiplied by the perpendicular distance R, it being a principle of geometry that a radius is perpendicular to a line drawn tangent to a circle, at the point of tangency. Also the opposing moment of W is $W \times r$. Hence, by the principle of moments,

$$F \times R = W \times r$$

Now, for comparison, we will apply the principle of work. Assuming this principle to be true, force F multiplied by the distance traversed by this force or by a given point on the rim of the large pulley, should equal the resistance W multiplied by the distance that the load is raised. In one revolution force F passes through a distance equal to the circumference of the pulley which is equal to $2 \times 3.1416 \times R = 6.2832 \times R$, and the hoisting rope passes through a distance equal to $2 \times 3.1416 \times r$. Hence, by the principle of work,

$$6.2832 \times F \times R = 6.2832 \times W \times r$$

The statement simply shows that $F \times R$ multiplied by 6.2832 equals $W \times r$ multiplied by the same number, and it is evident therefore, that the equality will not be altered by canceling the 6.2832 and writing

$$F \times R = W \times r$$

But this is the same statement obtained by applying the principle of moments; hence, we see that the principle of moments and the principle of work harmonize.

The basis of operation of a train of wheels is a continuation of the principle of work. For example, in the gear train represented by the diagram at the bottom of Handbook page 149, the continued product of the applied force F and the radii of the driven wheels equals the continued product of the resistance W and the radii of the drivers. In calculations, the pitch diameters, or the numbers of teeth in gear wheels may be used instead of the radii.

Efficiency of a Machine or Mechanism.—The efficiency of a machine is the ratio of the power delivered by the machine to the power received by it. For example, the efficiency of an electric motor is the ratio between the power delivered by the motor to the machinery which it drives, and the power it receives from the generator. Assume, for example, that a motor receives 50 kilowatts from the generator, but that the output of the motor is only 47 kilowatts. Then, the efficiency of the motor is $47 \div 50 = 94$ per cent. The efficiency of a machine tool is the ratio of the power consumed at the cutting tool to the power delivered by the driving belt. The efficiency of gearing is the ratio between the power obtained from the driven shaft to the power used by the driving shaft. Generally speaking, the efficiency of any machine or mechanism is the ratio of the "output" of power to the "input." The percentage of power representing the difference between the "input" and "output" has been dissipated through frictional and other mechanical losses.

Mechanical Efficiency.—If E represents the energy which a machine transforms into useful work or delivers at the driven end; L equals the energy loss through friction or dissipated in other ways; then

$$\text{Mechanical Efficiency} = \frac{E}{E + L}$$

In this case the total energy $E + L$ is assumed to be the amount that is transformed into useful and useless work. The actual total amount of energy, however, may be considerably larger than the amount represented by $E + L$. For example, in a steam engine there are heat losses due to radiation and steam condensation, and considerable heat energy supplied to an internal combustion engine is dissipated either through the cooling water or direct to the atmosphere. In other classes of mechanical and electrical machinery the total energy is much larger than that represented by the amount transformed into useful and useless work.

Absolute Efficiency.—If E_1 equals the full amount of energy or the true total, then

$$\text{Absolute Efficiency} = \frac{E}{E_1}$$

It is evident that absolute efficiency of a prime mover, such as a steam or gas engine, will be much lower than the mechanical efficiency. Ordinarily, the term efficiency as applied to engines and other classes of machinery. means the mechanical efficiency. The mechanical efficiency of reciprocating steam engines may vary from 85 to 95 per cent, but the *thermal* efficiency may range from 5 to 25 per cent, the smaller figure representing non-condensing engines of the cheaper class and the higher figure the best types.

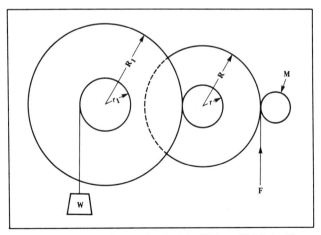

Fig. 4.   Determining the Power Required for Lifting a Weight by Means of Motor and Compound Train of Gearing

Example 6:—Assume that a motor driving through a compound train of gearing (see diagram, Fig. 4) is to lift a weight W of 1000 pounds. The pitch radius R = 6 inches; R_1 = 8 inches; pitch radius of pinion r = 2 inches; and radius of winding drum r_1 = 2½ inches. What motor horsepower will be required if the frictional loss in the gear train and bearings is assumed to be 10 per cent? The pitch-line velocity of the motor pinion M is 1200 feet per minute.

The problem is to determine first the tangential force F required at the pitch line of the motor pinion; then the equivalent horsepower is easily found. According to the formula at the bottom of Handbook page 149, which does not take into account frictional losses

$$F = \frac{1000 \times 2 \times 2½}{6 \times 8} = 104 \text{ pounds}$$

The pitch-line velocity of the motor pinion is 1200 feet per minute and as the friction loss is assumed to be 10 per cent, the mechanical efficiency equals 90 ÷ (90 + 10) = 0.90 or 90 per cent as commonly written; hence

$$\text{Horsepower} = \frac{104 \times 1200}{33,000 \times 0.90} = 4¼ \text{ approximately}$$

Example 7:—In actual designing practice, a 5 horsepower motor, or one of larger size, might be selected for the drive referred to in Example 6 (depending upon conditions), to provide extra power in case it is needed. Assume, however, in order to illustrate the procedure, that the gear train is to be modified so that the calculated horsepower will be 4 instead of 4¼; conditions otherwise are the same as in Example 6.

$$F = \frac{33,000 \times 4}{1200} = 110 \text{ pounds}$$

Hence since W = 1000 pounds,

$$1000 = \frac{110 \times 0.90 \times R \times R_1}{r \times r_1}$$

Insert any values for the pitch radii R, R_1, etc., that will balance the equation, so that the right-hand side equals 1000, at least approximately. Several trial solutions may be necessary to obtain a total of about 1000 and at the same time, secure properly proportioned gears that meet other requirements of the design. In this case, suppose the same radii are used, except R_1, which is increased from 8 to 8½ inches. Then

$$\frac{110 \times 0.90 \times 6 \times 8\frac{1}{2}}{2 \times 2\frac{1}{2}} = 1000 \text{ approximately}$$

This example shows that the increase in the radius of the last driven gear from 8 to 8½ inches makes it possible to use the four horsepower motor. The hoisting speed has been decreased somewhat and the center distance between the gears has been increased. These changes might or might not be objectionable in actual designing pratice, depending upon the particular requirements.

Force Required to Turn a Screw Used for Elevating or Lowering Loads.—In determining the force which must be applied at the end of a given lever-arm in order to turn a screw (or nut surrounding it), there are two conditions to be considered: (1) When rotation is such that the load *resists* the movement of the screw, as in raising a load with a screw jack; (2) when rotation is such that the load *assists* the movement of the screw, as in lowering a load. The formulas at the bottom of the table on Handbook page 150, apply to both of these conditions. When the load resists the screw movement, use the formula "for motion in a direction opposite to Q." When the load assists the screw movement, use the formula "for motion in the same direction as Q."

If the lead of the thread is large in proportion to the diameter so that the helix angle is large, the force F may have a negative value, which indicates that the screw will turn due to the load alone, unless resisted by a force which is great enough to prevent rotation of a non-locking screw.

Example 8:—A screw is to be used for elevating a load Q of 6000 pounds. The pitch diameter is 4 inches, the lead is 0.75 inch and the coefficient of friction μ between screw and nut is assumed to be 0.150. What force F will be required at the end of a lever arm R of 10 inches? In this example, the load is in the direction opposite to arrow Q (see diagram at bottom of the table on Handbook page 150).

$$F = 6000 \times \frac{0.75 + 6.2832 \times 0.150 \times 2}{6.2832 \times 2 - 0.150 \times 0.75} \times \frac{2}{10} = 254 \text{ pounds}$$

Example 9:—What force *F* will be required to lower a load of 6000 pounds using the screw referred to in Example 8? In this case, the load assists in turning the screw; hence

$$F = 6000 \times \frac{6.2832 \times 0.150 \times 2 - 0.75}{6.2832 \times 2 + 0.150 \times 0.75} \times \frac{2}{10} = 107 \text{ pounds}$$

Coefficients of Friction for Screws and Their Efficiency.—According to experiments by Professor Kingsbury made with square-threaded screws, a friction coefficient of 0.10 is about right for pressures less than 3000 pounds per square inch and velocities above 50 feet per minute, assuming that fair lubrication is maintained. If the pressures vary from 3000 to 10,000 pounds per square inch, a coefficient of 0.15 is recommended for low velocities. The coefficient of friction varies with lubrication and the materials used for the screw and nut. For pressures of 3000 pounds per square inch and using heavy machinery oil as a lubricant, the coefficients were as follows: Mild steel screw and cast-iron nut, 0.132; mild steel nut, 0.147; cast brass nut, 0.127. For pressures of 10,000 pounds per square inch using a mild steel screw, the coefficients were, for a cast-iron nut, 0.136; for a mild steel nut, 0.141; for a cast brass nut, 0.136. For dry screws, the coefficient may be 0.3 to 0.4 or higher.

Frictional resistance is proportional to the normal pressure, and for a thread of angular form, the increase in the coefficient of friction is equivalent practically to $\mu \sec \beta$, in which β equals one-half the included thread angle; hence, for a sixty-degree thread, a coefficient of 1.155μ may be used. The square form of thread has a somewhat higher efficiency than threads with sloping sides, although when the angle of the thread form is comparatively small, as in the case of an Acme thread, there is little increase in frictional losses. Multiple-thread screws are much more efficient than single-thread screws, as the efficiency is affected by the helix angle of the thread.

The efficiency between a screw and nut increases quite rapidly for helix angles up to 10 or 15 degrees (measured from a plane perpendicular to the screw axis). The efficiency remains nearly constant for angles between about 25 and 65 degrees, and the angle of maximum efficiency is between 40 and 50 degrees. A screw will not be self-locking if the efficiency exceeds 50 per cent. For example, the screw of a jack or other lifting or hoisting appliance would turn under the action of the load if the efficiency were over 50 per cent. It is evident that maximum efficiency for power transmission screws often is impractical, as for example, when the smaller helix angles are required to permit moving a given load by the application of a smaller force or turning moment than would be needed for a multiple screw thread.

In determining the efficiency of a screw and a nut, the helix angle of the thread and the coefficient of friction are the important factors. If E equals the efficiency, A equals the helix angle, measured from a plane perpendicular to the screw axis, and μ equals the coefficient of friction between the screw thread and nut, then the efficiency may be determined by the following formula, which does not take into account any additional friction losses, such as may occur between a thrust collar and its bearing surfaces:

$$E = \frac{\tan A(1 - \mu \tan A)}{\tan A + \mu}$$

This formula would be suitable for a screw having ball-bearing thrust collars. Where collar friction should be taken into account, a fair approximation may be obtained by changing the denominator of the foregoing formula to $\tan A + 2\mu$. Otherwise the formula remains the same.

Angles and Angular Velocity Expressed in Radians.—There are three systems generally used to indicate the sizes of angles, which are ordinarily measured by the number of degrees in the arc subtended by the sides of the angle. Thus if the arc subtended by the sides of the angle equals one-sixth of the circumference, the angle is said to be 60 degrees. Angles are also designated as multiples of a right angle. As an example of this, the sum of the interior angles of any polygon equals the number of sides less two, times two right angles. Thus the sum of the interior angles of an octagon equals $(8 - 2) \times 2 \times 90 = 6 \times 180 = 1080$ degrees. Hence each interior angle equals $1080 \div 8 = 135$ degrees.

A third method of designating the size of an angle is very helpful in certain problems. This method makes use of radians. A radian is defined as a central angle, the subtended arc of which equals the radius of the arc.

Using the symbols on Handbook page 200, v may represent the length of an arc as well as the velocity of a point on the periphery of a body. Then according to the definition of a radian: $\omega = v/r$ or, the angle in radians equals the length of the arc divided by the radius. Both the length of the arc and the radius must, of course, have the same unit of measurement—both must be in feet or inches or centimeters, etc. By rearranging the preceding equation:

$$v = \omega r \quad \text{and} \quad r = \frac{v}{\omega}$$

These three formulas will solve practically every problem involving radians.

The circumference of a circle equals πd or $2\pi r$ which equals $6.2832r$. This indicates that a radius is contained in a circumference 6.2832 times; hence there are 6.2832 radians in a circumference. Since a circumference represents 360 degrees, one radian equals $360 \div 6.2832 = 57.2958$ degrees. Since 57.2958 degrees $= 1$ radian, 1 degree $= 1$ radian $\div 57.2958 = 0.01745$ radian.

Example 10:—2.5 radians equal how many degrees? One radian $= 57.2958$ degrees; hence, 2.5 radians $= 57.2958 \times 2.5 = 143.239$ degrees.

Example 11:—22° 31′ 12″ = how many radians? 12 seconds $= ^{12}\!/_{60} = ^1\!/_5 = 0.2$ minute; $31.2′ \div 60 = 0.52$ degree. One radian $= 57.3$ degrees approximately. $22.52° = 22.52 \div 57.3 = 0.393$ radian.

Example 12:—In the figure on Handbook page 70, Let $l = v = 30$ inches; $r = 50$ inches; find the central angle. $\omega = v/r = ^{30}\!/_{50} = ^3\!/_5 = 0.6$ radian.

$$57.2958 \times 0.6 = 34° \; 22.6′$$

Example 13:— $3\pi/4$ radians equal how many degrees? 2π radians $= 360°$; π radians $= 180°$. $3\pi/4 = ^3\!/_4 \times 180 = 135$ degrees

Example 14:—A 20-inch grinding wheel has a surface speed of 6000 feet per minute. What is the angular velocity?

The radius $(r) = ^{10}\!/_{12}$ feet; the velocity (v) in feet per second $= ^{6000}\!/_{60}$; hence,

$$\omega = \frac{6000}{60 \times ^{10}\!/_{12}} = 120 \text{ radians per second}$$

Example 15:—Use table on page 127 to solve Example 11.

20° = 0.349066 radians	
2° = 0.034907	″
31′ = 0.009018	″
12″ = 0.000058	″
22° 31′ 12″ = 0.393049	″

Example 16:—7.23 radians equals how many degrees? On Handbook page 128 find:

7.0	radians	=	401°	4′	14″
0.2	″	=	11°	27′	33″
0.03	″	=	1°	43′	8″
7.23	″	=	414°	14′	55″

PRACTICE EXERCISES FOR SECTION 14
For answers to all practice exercise problems or questions
see Section 22

1. In what respect does a foot-pound differ from a pound?

2. If 100 pounds is dropped, how much energy will it be capable of exerting after falling 10 feet?

3. Can the force of a hammer-blow be expressed in pounds?

4. If a 2-pound hammer is moving 30 feet per second, what is its kinetic energy?

5. If the hammer referred to in question 4 drives a nail into a board ¼ inch, what is the average force of the blow?

6. What relationship is there between the muzzle velocity of a projectile fired upward and the velocity with which the projectile strikes the ground?

7. What is the difference between the composition of forces and the resolution of forces?

8. If four equal forces act along lines 90 degrees apart through a given point, what is the shape of the corresponding polygon of forces?

9. Skids are to be employed for transferring boxed machinery from one floor to the floor above. If these skids are inclined at an angle of 35 degrees, what force in pounds, applied parallel to the skids, will be required to slide a boxed machine weighing 2500 pounds up the incline, assuming that the coefficient of friction is 0.20?

10. Referring to question 9, if the force or pull were applied in a horizontal direction instead of in line with the skids, what increase, if any, would be required?

11. Will the boxed machine referred to in question 9 slide down the skids by gravity?

12. At what angle will the skids require to be before the boxed machine referred to in question 9 begins to slide by gravity?

13. What name is applied to the angle which marks the dividing line between sliding and non-sliding when a body is placed on an inclined plane?

14. How is the "angle of repose" determined?

15. What figure or value is commonly used in engineering calculations for acceleration due to gravity?

16. Is the value commonly used for acceleration due to gravity strictly accurate for any locality?

17. A flywheel 3 feet in diameter has a rim speed of 1200 feet per minute, and another flywheel 6 feet in diameter has the same rim speed. Will the rim stress or the force tending to burst the larger flywheel be greater than the force in the rim of the smaller flywheel?

18. What factors of safety are commonly used in designing flywheels?

19. Does the stress in the rim of a flywheel increase in proportion to the rim velocity?

20. What is generally considered the maximum safe speed for the rim of a solid or one-piece cast-iron flywheel?

21. Why is a well-constructed wood flywheel adapted to higher speeds than one made of cast iron?

22. What is the meaning of the term "critical speed" as applied to a rotating body?

23. How is angular velocity generally expressed?

24. What is a radian and how is its angle indicated?

25. How many degrees are there in 2.82 radians?

26. How many degrees are in the following radians: $\pi/3$; $2\pi/5$; 2π?

27. Reduce to radians: 63°; 45° 32'; 6° 37' 46"; 22° 22' 22".

28. Find the angular velocity in radians per second of the following: 157 r.p.m.; 275 r.p.m.; 324 r.p.m.

29. Why do the values in the *l* column on Handbook pages 70 and 71 equal those in the radian column on page 127?

30. If the length of the arc of a sector is 4⅞ inches and the radius is 6⅞ inches, find the central angle.

31. A 12-inch grinding wheel has a surface speed of a mile a minute. Find its angular velocity and its revolutions per minute.

32. The radius of a circle is 1½ inches and the central angle is 60 degrees. Find the length of the arc.

33. If an angle of 34° 12′ subtends of arc of 16.25 inches, find the radius of the arc.

SECTION 15

STRENGTH OF MATERIALS

HANDBOOK Pages 203–314

The Strength of Materials section of MACHINERY'S HANDBOOK contains fundamental formulas and data for use in proportioning parts that are common to almost every type of machine or mechanical structure. In designing machine parts, factors other than strength often are of vital importance. For example, some parts are made much larger than required for strength alone to resist extreme vibrations, or deflection, or wear; consequently, many machine parts cannot be designed merely by mathematical or strength calculations and their proportions should, if possible, be based upon experience or upon similar designs that have proved successful. It is evident that no engineering handbook can take into account the endless variety of requirements relating to all types of mechanical apparatus and it is necessary for the designer to determine these local requirements for himself; but, even when the strength factor is secondary due to some other requirement, the strength, especially of the more important parts, should be calculated, in many instances, merely to prove that it will be sufficient.

In designing for strength, the part is so proportioned that the maximum working stress likely to be encountered will not exceed the strength of the material by a suitable margin. The design is accomplished by the use of a factor of safety. The relationship between the working stress s_w, the strength of the material, S_m, and the factor of safety, f_s, is given by Equation (1) on page 205 of the Handbook:

$$s_w = \frac{S_m}{f_s} \qquad (a)$$

The value selected for the strength of the material, S_m, depends on the type of material, whether failure is expected to occur because of tensile, compressive, or shear stress, and on whether the stresses are constant, fluctuating, or are abruptly applied as in the case of shock loading. In general the value of S_m is based on yield strength for ductile materials, ultimate strength for brittle materials, and fatigue strength for parts subject to cyclic stresses. Moreover, the value for S_m must be for the temperature at which the part operates. Values of S_m for common materials at 68° F can be obtained from the tables in MACHINERY'S HANDBOOK from pages 512 to 513. Factors from the table given on Handbook page 514 "Influence of Temperature on the Strength of Metals" can be used to convert strength values at 68° F to values applicable at elevated temperatures. For heat treated carbon and alloy steel parts, see data on pages 460 and 465.

The factor of safety depends on the relative importance of reliability, weight, and cost. General recommendations are given in the Handbook on pages 205 and 206.

Working stress is dependent on the shape of the part, hence on a stress concentration factor, and on a nominal stress associated with the way in which the part is loaded. Equations and data for calculating nominal stresses, stress concentration factors, and working stresses are given on pages 206 to 223.

Example 1:—Determine the allowable working stress for a part that is to be made from SAE 1112 free cutting steel; the part is loaded in such a way that failure is expected to occur in tension when the yield strength has been exceeded. A factor of safety of 3 is to be used.

From the table, "Strength Data for Iron and Steel" on page 512 of the Handbook, a value of 30,000 psi is selected for the strength of the material, S_m. Working stress s_w is calculated from Equation (a) as follows:

$$s_w = \frac{30,000}{3} = 10,000 \text{ psi}$$

Finding Diameter of Bar to Resist Safely in a Given Load.—Assume that a direct tension load, F, is applied to a bar such that the force acts along the longitudinal axis of the bar. From page 214 of the Handbook, the following equation is given for calculating the nominal stress:

$$\sigma = \frac{F}{A} \tag{b}$$

where A is the cross-sectional area of the bar. Equation (2) on Handbook page 206 relates the nominal stress to the stress concentration factor, K, and working stress, s_w:

$$s_w = K\sigma \tag{c}$$

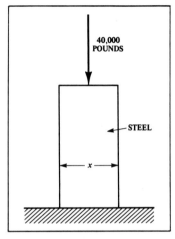

Fig. 1. Calculating Diameter *x* to Support a Given Load Safely

Combining Equations (a), (b), and (c), results in the following:

$$\frac{S_m}{Kf_s} = \frac{F}{A} \tag{d}$$

Example 2:—A structural steel bar supports in tension a load of 40,000 pounds. The load is gradually applied, and then after having reached its maximum value, is gradually removed. Find the diameter of round bar required.

According to the table on Handbook page 512, the yield strength of structural steel is 33,000 psi. Suppose that a factor of safety of 3 is used and a stress concentration factor of 1.1. Then, inserting known values in Equation (d):

$$\frac{33,000}{1.1 \times 3} = \frac{40,000}{A}; \; A = \frac{40,000 \times 3.3}{33,000}; \; A = 4 \text{ square inches}$$

Hence, the cross-section of the bar must be about 4 square inches. As the bar is circular in section, the diameter must then be about 2¼ inches.

Diameter of Bar to Resist Compression.—If a short bar is subjected to compression in such a way that the line of application of the load coincides with the longitudinal axis of the bar, the formula for nominal stress is the same as for direct tension loading; Equation (b) and hence Equation (d) may be also applied to direct compression loading.

Example 3:—A short structural steel bar supports in compression a load of 40,000 pounds. (See Fig. 1.) The load is steady. Find the diameter of the bar required.

From page 512 in the Handbook, the yield strength of structural steel is 33,000 psi. If a stress concentration factor of 1.1 and a factor of safety of 2.5 are used, then substituting values into Equation (d):

$$\frac{33,000}{1.1 \times 2.5} = \frac{40,000}{A}; A = 3.33 \text{ square inches}$$

The diameter of a bar, the cross-section of which is 3.33 square inches is 2 1/16 inches approximately.

According to a general rule, the simple formulas that apply to compression should be used only if the length of the member being compressed is not greater than 6 times the least cross-sectional dimension. For example, these formulas should be applied to round bars only when the length of the bar is less than 6 times the diameter. If the bar is rectangular, the formulas should be applied only to bars having a length less than 6 times the shortest side of the rectangle. When bars are longer than this, a compressive stress causes a sidewise bending action, and an even distribution of the compression stresses over the total area of the cross-section is no longer to be depended upon. Special formulas for long bars or columns will be found on Handbook page 290; see also text beginning on page 288, "Strength of Columns or Struts."

Diameter of Pin to Resist Shearing Stress.—The pin E shown in the illustration, Fig. 2, is subjected to shear. Parts G and B are held togther by the pin and tend to shear it off at C and D. The areas resisting the shearing action are equal to the cross-sectional areas of the pin at these points.

From the "Table of Simple Stresses" on page 214 of the Handbook, the equation for direct shear is:

$$\tau = \frac{F}{A} \tag{e}$$

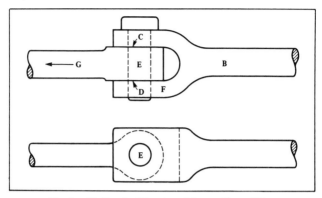

**Fig. 2.   Finding the Diameter of Connecting-rod Pin
to Resist a Known Load G**

τ is a simple stress related to the working stress, s_w, by Equation
(3) on Handbook page 208:

$$s_w = K\tau \qquad\qquad (f)$$

where K is a stress concentration factor. Combining Equations (a),
(e), and (f) gives Equation (d) on page 122, where S_m is of course
the shearing strength of the material.

If a pin is subjected to shear as in Fig. 2, so that two surfaces,
as at C and D, must fail by shearing before breakage occurs, the
areas of both surfaces must be taken into consideration when calculating the strength. The pin is then said to be in *double shear*. If
the lower part F of connecting-rod B were removed, so that member
G were connected with B by a pin subjected to shear at C only, the
pin would be said to be in *single shear*.

Example 4:—Assume that in Fig. 2 the load at G pulling on the
connecting-rod is 20,000 pounds. The material of the pin is SAE
1025 steel. The load is applied in such a manner that shocks are
liable to occur. Find the required dimensions for the pin.

Since the pins are subject to shock loading, the nominal stress
resulting from the application of the 20,000 pound load must be assumed to be twice as great (see Handbook page 284) as it would be

if the load were gradually applied or steady. From page 512 in the Handbook, the ultimate strength in shear for SAE 1025 steel is 75 per cent of 60,000 or 45,000 psi. A factor of safety of 3 and a stress concentration factor of 1.8 are to be used. Substituting values into Equation (d):

$$\frac{45,000}{1.8 \times 3} = \frac{2 \times 20,000}{A}; A = \frac{10.8 \times 20,000}{45,000}$$

$$A = 4.8 \text{ sq. ins.}$$

As the pin is in double shear, that is, as there are two surfaces C and D over which the shearing stress is distributed, each of them must have an area of one-half the total shearing area A. In this case, then, the cross-sectional area of the pin will be 2.4 square inches, and the diameter of the pin, to give a cross-sectional area of 2.4 square inches, must be 1¾ inches.

Beams and Stresses to Which They are Subjected.—Parts of machines and structures subjected to bending are known mechanically as *beams*. Hence, in this sense, a lever fixed at one end and subjected to a force at its other end, a rod supported at both ends and subjected to a load at its center, or the overhanging arm of a jib crane, would all be known as beams.

The stresses in a beam are principally tension and compression stresses. If a beam is supported at the ends and a load rests upon the upper side, the lower fibers will be stretched by the bending action and will be subjected to a tensile stress, while the upper fibers will be compressed and be subjected to a compressive stress. There will be a slight lengthening of the fibers in the lower part of the beam, while those on the upper side will be somewhat shorter, depending upon the amount of deflection. If we assume that the beam is either round or square in cross-section, there will be a layer or surface through its center line which will be neither in compression nor in tension. This surface is known as the neutral surface. The stresses of the individual layers or fibers of the beam will be proportional to their distances from the neutral surface, the stresses being greater the farther away from the neutral surface the fiber is located. Hence, there is no stress on the fibers in the neutral surface, but there is a maximum tension on the fibers at the extreme lower side and a maximum compression on the fibers at the extreme upper side of the beam. In calculating the strength of beams, it is, therefore, necessary only to determine that the fibers of the beam which are at the greatest distance from the neutral surface are not stressed beyond the safe working stress of the material. If this is the case, all the other parts of the section of the beam are not stressed beyond the safe working stress of the material.

In addition to the tension and compression stresses, a loaded beam is also subjected to a stress which tends to shear it. This shearing stress depends upon the magnitude and kind of load. In most cases, the shearing action can be ignored for metal beams, especially if the beams are long and the loads far from the supports. If the beams are very short and the load quite close to a support, then the shearing stress may become equal to or greater than the tension or compression stresses in the beam and the beam should then be calculated for shear.

Beam Formulas.—The bending action of a load upon a beam is called the *bending moment*. For example, in Fig. 3 the load P acting downward on the free end of the cantilever beam has a moment or bending action about the support at A equal to the load multiplied by its distance from the support. The bending moment is commonly expressed in inch-pounds, the load being expressed in pounds and the lever arm or distance from the support in inches. The length of the lever arm should always be measured in a direction at right angles to the direction of the load. Thus, in Fig. 4, the bending moment is not $P \times a$, but is $P \times l$, because l is measured in a direction at right angles to the direction of the load P.

The property of a beam to resist the bending action or the bending moment is called the *moment of resistance* of the beam. It is evident that the bending moment must be equal to the moment of resistance.

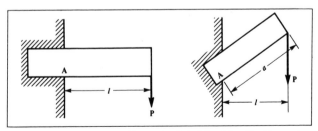

Figs. 3 and 4. Diagrams Illustrating Principle of Bending Moments

The moment of resistance, in turn, is equal to the stress in the fiber farthest away from the neutral plane multiplied by the section modulus. The *section modulus* is a factor which depends upon the shape and size of the cross-section of a beam, and is given for different cross-sections in all engineering handbooks. (See table "Moments of Inertia, Section Moduli, etc., of Sections" in the Handbook pages 226 to 235.) The section modulus, in turn, equals the *moment of inertia* of the cross-section, divided by the distance from the neutral surface to the most extreme fiber. The moment of inertia formulas for various cross-sections will also be found in the table just referred to.

The following formula on Handbook page 214 may be given as the fundamental formula for bending of beams:

$$\sigma = \pm \frac{M}{Z} = \pm \frac{My}{I} \qquad (g)$$

The moment of inertia I is a property of the cross-section that determines its relative strength. In calculations of strength of materials, a handbook is necessary because of the tabulated formulas and data relating to section moduli and moments of inertia, areas of cross-sections, etc., to be found therein.

There are many different ways in which a beam can be supported and loaded, and the bending moment caused by a given load varies greatly according to whether the beam is supported at one end only or at both ends, and also whether it is freely supported at the ends or is held firmly. Then the load may be equally distributed over the full length of the beam or may be applied at one point either in the center or near to one or the other of the supports. The point where the stress is maximum is generally called the critical point. The stress at the critical point equals bending moment divided by section modulus. Formulas for determining the stresses at the critical points will be found in the table of beam formulas, Handbook pages 260 to 271.

Example 5:—A rectangular steel bar 2 inches thick and firmly built into a wall, as shown in Fig. 5, is to support 3000 pounds at its outer end 36 inches from the wall. What would be the necessary depth h of the beam to support this weight safely?

The bending moment equals the load times the distance from the point of support, or, 3000 × 36 = 108,000 inch-pounds.

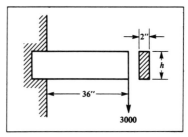

Fig. 5. Determining the Depth _h_ of a Beam to Support a Known Weight

By combining Equations (a), (c), and (g), the following equation is obtained:

$$\frac{S_m}{Kf_s} = \frac{M}{Z} \tag{h}$$

If the beam is made from structural steel, the value for S_m, based on yield strength, from page 512 in the Handbook is 33,000 psi. Using a stress concentration factor of 1.1 and a factor of safety of 2.5, values may be inserted into the above equation:

$$\frac{33,000}{1.1 \times 2.5} = \frac{108,000}{Z}; \ Z = \frac{2.75 \times 108,000}{33,000}; \ Z = 9 \text{ inches}^3$$

The section modulus for a rectangle equals $bd^2/6$, in which b is the length of the shorter side and d of the longer side of the rectangle (see Handbook page 260); hence, $Z = bd^2/6$.

But $Z = 9$ and $b = 2$. Inserting these values into the formula, we have:

$$9 = \frac{2d^2}{6}$$

from which $d^2 = 27$, and $d = 5.2$ inches. This value d corresponds to dimension h in Fig. 5. Hence, the required depth of the beam to support a load of 3000 pounds at the outer end with a factor of safety of 3 would be 5.2 inches.

In calculating beams having either rectangular or circular cross-sections, the formulas on Handbook pages 274 and 275 will be found convenient to use. A beam loaded as shown by Fig. 5 is similar to the first diagram on Handbook page 274. If the formula on this page for determining height h is applied to Example 5, Fig. 5, then

$$h = \sqrt{\frac{6lW}{bf}} = \sqrt{\frac{6 \times 36 \times 3000}{2 \times 12,000}} = 5.2 \text{ inches}$$

In the above calculation the stress value f is equivalent to S_m/Kf_s.

Example 6:—A steel I-beam is to be used as a crane trolley track. This I-beam is to be supported at the ends and the unsupported span is 20 feet long. The maximum load is 6000 pounds and the nominal stress is not to exceed 10,000 pounds per square inch. Determine the size of standard I-beam; also the maximum deflection when the load is at the center of the beam.

The foregoing conditions are represented by Case 2, Handbook page 260. A formula for the stress at the critical point is $Wl/4Z$. As explained on page 259, all dimensions are in inches, and the minus sign preceding a formula merely denotes compression of the upper fibers and tension in the lower fibers.

Inserting the known values in the formula:

$$10,000 = \frac{6000 \times 240}{4Z}; \text{ hence}$$
$$Z = \frac{6000 \times 240}{10,000 \times 4} = 36$$

The table of standard I-beams on Handbook page 253 shows that a 12-inch I-beam which weighs 31.8 pounds per foot has a section modulus of 36.4.

The formula for maximum deflection (see Handbook page 261 Case 2) is $Wl^3/48EI$. According to the table on Handbook page 512, the modulus of elasticity (E) of structural steel is 29,000,000. As Z = moment of inertia I ÷ distance from neutral axis to extreme fiber (see Handbook page 259), then for a 12-inch I-beam $I = 6Z = 216$; hence,

$$\text{Maximum deflection} = \frac{6000 \times 240^3}{48 \times 29,000,000 \times 216} = 0.27 \text{ inch}$$

Example 7:—All conditions are the same as in Example 6, excepting that the maximum deflection at the "critical point" or center of the I-beam, must not exceed ⅛ inch. What size I-beam is required?

To meet the requirement regarding deflection

$$\frac{1}{8} = \frac{Wl^3}{48EI}; \quad \text{therefore}$$

$$I = \frac{8Wl^3}{48E} = \frac{8 \times 6000 \times 240^3}{48 \times 29,000,000} = 476$$

If x = distance from neutral axis to most remote fiber (½ beam depth in this case) then $Z = I/x$ and the table on Handbook page 253 shows that a 15-inch, 50-pound I-beam should be used because it has a section modulus of 64.8 and 476/7.5 = 63.5 nearly.

If 476 were divided by 6 (½ depth of a 12-inch I-beam) the result would be much higher than the section modulus of any standard 12-inch I-beam (476 ÷ 6 = 79.3); moreover, 476 ÷ 9 = 53 which shows that an 18-inch I-beam is larger than is necessary because the lightest beam of this size has a section modulus of 81.9.

Example 8:—If the speed of a motor is 1200 revolutions per minute and if its driving pinion has a pitch diameter of 3 inches, determine the torsional moment to which the pinion shaft is subjected, assuming that 10 horsepower is being transmitted.

If W = tangential load in pounds, H = the number of horsepower and V = pitch-line velocity in feet per minute.

$$W = \frac{33,000 \times H}{V}$$

$$= \frac{33,000 \times 10}{943}$$

$$= 350 \text{ pounds}$$

The torsional moment = W × pitch radius of pinion = 350 × 1.5 = 525 pound-inches (or inch-pounds).

Example 9:—If the pinion referred to in Example 8 drives a gear having a pitch diameter of 12 inches, to what torsional or turning moment is the gear shaft subjected?

The torque or torsional moment in any case = pitch radius of gear × tangential load. The latter is the same for both gear and pinion; hence, torsional moment of gear = 350 × 6 = 2100 inch-pounds.

The torsional moment or the turning effect of a force which tends to produce rotation depends (1) upon the magnitude of the force acting, and (2) upon the distance of the force from the axis of rotation, measuring along a line at right angles to the line of action of the force.

PRACTICE EXERCISES FOR SECTION 15

For answers to all practice exercise problems or questions
see Section 22

1. What is a "factor of safety" and why are different factors used in machine design?

2. If the ultimate strength of a steel rod is 60,000 pounds per square inch and the factor of safety is 5, what is the equivalent working stress?

3. If a steel bar must withstand a maximum pull of 9000 pounds and if the maximum nominal stress must not exceed 12,000 pounds per square inch, what diameter bar is required?

4. Is a steel rod stronger when at ordinary room temperature or when heated, say, to 500° F?

5. What is the meaning of the term "elastic limit"?

6. Approximately what percentages of copper and zinc in brass result in the greatest tensile strength?

7. If four 10-foot long pipes are to be used to support a water tank installation weighing 100,000 pounds, what diameter standard weight pipe is required?

SECTION 16

DESIGN OF SHAFTS AND KEYS FOR POWER TRANSMISSION

HANDBOOK Pages 286–314

This section is a review of the general procedure in designing shafts to resist torsional and also combined torsional and bending stresses. The diameter of a shaft through which power is transmitted depends, for a given shaft material, upon the amount and kind of stress or stresses to which the shaft is subjected. In order to illustrate the general procedure, we shall assume first that the shaft is subjected only to a uniform torsional or twisting stress and that there is no additional bending stress which needs to be considered in determining the diameter.

Example 1:—A lineshaft carrying pulleys located close to the bearings is to transmit 50 horsepower at 1200 revolutions per minute. If the load is gradually applied, and steady, what diameter steel shaft is required assuming that the pulleys are fastened to the shaft by means of keys and that the bending stresses caused by the pull of the belts are negligible?

According to the former American Standard Association's Code for the Design of Transmission Shafting, the diameter of shaft required to meet the stated conditions can be determined by using the following formula (Formula 1b on page 310 of the Handbook):

$$D = B \sqrt[3]{\frac{321,000\ K_t P}{S_s N}}$$

In this formula, D = required shaft diameter in inches; B = a factor, which for solid shafts is taken as 1; K_t = combined shock and fatigue factor; P = maximum horsepower transmitted by shaft; S_s = maximum allowable torsional shearing stress in pounds per square inch; and N = shaft speed in revolutions per minute.

From Table 1 on page 311 of the Handbook, K_t = 1.0 for gradually applied and steady loads, and from Table 2 the recommended maximum allowable working stress for "Commercial Steel" shafting with keyways subjected to pure torsion loads is 6000 pounds per square inch. Substituting in the formula,

$$D = 1 \times \sqrt[3]{\frac{321,000 \times 1.0 \times 50}{6000 \times 1200}} = 1.306 \text{ inches}$$

The nearest standard size transmission shafting from the table on Handbook page 310 is $1\frac{7}{16}$ inches.

Example 2:—If in Example 1 the shaft diameter had been determined by using Formula 5b on Handbook page 302, what would the result have been and why?

$$D = \sqrt[3]{\frac{53.5P}{N}} = \sqrt[3]{\frac{53.5 \times 50}{1200}} = 1.306 \text{ inches}$$

This formula gives the same shaft diameter as was previously determined because it is a simplified form of the first formula used and contains the same values of K_t and S_s, but combined as the single constant 53.5. For lineshafts carrying pulleys under conditions ordinarily encountered, this simplified formula is usually quite satisfactory; but, where conditions of shock loading are known to exist, it is safer to use Formula 1b on Handbook page 310 which takes such conditions into account.

Shafts Subjected to Combined Stresses.—The preceding formulas are based on the assumption that the shaft is subjected to torsional stresses only. Many shafts, however, must withstand stresses that result from combinations of torsion, bending, and shock loading. In such cases it is necessary to use formulas which take such conditions into account.

Example 3:—Suppose that after the lineshaft in Example 1 was installed it became necessary to relocate a machine which was being driven by one of the pulleys on the shaft. Because of the new machine location it was necessary to move the pulley on the lineshaft farther away from the nearest bearing and, as a result, a bending moment of 2000 inch-pounds has been introduced. Is the $1\frac{7}{16}$-inch diameter shaft sufficient to take this additional stress or will it be necessary to relocate the bearing to provide better support?

Since the shaft now has both bending and torsional loads acting on it, Formula 3b on Handbook page 310 should be used to compute the required shaft diameter. This diameter is then compared with the 1⁷⁄₁₆-inch diameter previously determined.

$$D = B \sqrt[3]{\frac{5.1}{p_t} \sqrt{(K_m M)^2 + \left(\frac{63,000\ K_t P}{N}\right)^2}}$$

In this formula B, K_t, P, and N are quantities previously defined and p_t = maximum allowable shearing stress under combined loading conditions in pounds per square inch; K_m = combined shock and fatigue factor; and M = maximum bending moment in inch-pounds.

From Table 1 on Handbook page 311, K_m = 1.5 for gradually applied and steady loads and from Table 2, p_t = 6000 pounds per square inch. Substituting in the formula,

$$D = 1 \times \sqrt[3]{\frac{5.1}{6000} \sqrt{(1.5 \times 2000)^2 + \left(\frac{63,000 \times 1 \times 50}{1200}\right)^2}}$$

$$= \sqrt[3]{\frac{5.1}{6000} \sqrt{9,000,000 + 6,890,625}} = \sqrt[3]{\frac{5.1}{6000} \times 3986}$$

$$= \sqrt[3]{3.388} = 1.502 \text{ inches, say, } 1\tfrac{1}{2} \text{ inches}$$

Since this diameter is larger than the 1⁷⁄₁₆ diameter actually used for the shaft in Example 1, it will be necessary to relocate the bearing so that it is closer to the pulley, thus reducing the bending moment. The 1⁷⁄₁₆-inch diameter shaft will then be able to operate within the allowable working stress for which it was originally designed.

Design of Shafts to Resist Torsional Deflection.—In many cases shafts must be proportioned not only to provide the strength required to transmit a given torque, but also to prevent torsional deflection (twisting) through a greater angle than has been found satisfactory for a given type of service. This is particularly true in the case of machine shafts and machine-tool spindles.

For ordinary service it is customary that the angle of twist of machine shafts be limited to ¹⁄₁₀ degree per foot of shaft length, and for machine shafts subject to load reversals, ¹⁄₂₀ degree per foot of shaft length. As explained in the Handbook, the usual design procedure for shafting which is to have a specified maximum angular deflection is to compute the diameter of shaft required based on both deflection and strength considerations and then to choose the larger of the two diameters thus determined.

Example 4:—A 6-foot-long feed shaft is to transmit a torque of 200 inch-pounds. If there are no bending stresses and the shaft is to be limited to a torsional deflection of $\frac{1}{20}$ degree per foot of length, what diameter shaft should be used? The shaft is to be made of cold drawn steel and is to be designed for a maximum working stress of 6000 pounds per square inch in torsion.

The diameter of shaft required for a maximum angular deflection α is given by Formula 8 on Handbook page 304.

$$D = 4.9 \sqrt[4]{\frac{Tl}{G\alpha}}$$

In this formula T = applied torque in inch-pounds; l = length of shaft in inches; G = torsional modulus of elasticity which for steel is 11,500,000 pounds per square inch; and α = angular deflection of shaft in degrees.

In the problem at hand, T = 200 inch-pounds; l = 6 × 12 = 72 inches; and α = 6 × $\frac{1}{20}$ = 0.3 degrees.

$$D = 4.9 \sqrt[4]{\frac{200 \times 72}{11,500,000 \times 0.3}} = 4.9 \sqrt[4]{0.0041739}$$

$$= 4.9 \times 0.254 = 1.24 \text{ inches}$$

The diameter of the shaft based on strength considerations is obtained by using Formula 3a on Handbook page 302:

$$D = \sqrt[3]{\frac{5.1T}{S_s}} = \sqrt[3]{\frac{5.1 \times 200}{6000}} = \sqrt[3]{0.17} = 0.55 \text{ inch}$$

Therefore, since the diameter based on torsional deflection considerations is the larger of the two values obtained, the nearest standard diameter, $1\frac{1}{4}$ inches, should be used.

Selection of Key Size Based on Shaft Size.—Keys are generally proportioned in relation to shaft diameter instead of in relation to torsional load to be transmitted because of practical reasons such as standardization of keys and shafts. Thus, on Handbook page 2236 will be found a table entitled "Key Size Versus Shaft Diameter."

Dimensions of both square and rectangular keys are given, but for shaft diameters up to and including 6½ inches, square keys are preferred, while for larger shafts, rectangular keys are commonly used.

Two rules which base key length on shaft size are: (1) $L = 1.5\ D$ and (2) $L = 0.3\ D^2 \div T$, where L = length of key, D = diameter of shaft and T = key thickness.

If the keyseat is to have fillets and the key is to be chamfered, suggested dimensions for these modifications are given on Handbook page 2239. If a set screw is to be used over the key, suggested sizes are given in the table on Handbook page 2239.

Example 5:—If the maximum torque output of a 2-inch diameter shaft is to be transmitted to a keyed pulley, what should be the proportions of the key?

According to the table on Handbook page 2239, a ½-inch square key would be preferred. If a rectangular key were selected, its dimensions would be ½ by ⅜ inch. According to rule 1 its length would be 3 inches.

The key and keyseat may be so proportioned as to provide a clearance or an interference fit. The table on Handbook page 2238 gives tolerances for widths and depths of keys and keyseats to provide Class 1 (clearance) and Class 2 (interference) fits. An additional Class 3 (interference) fit which has not been standardized is mentioned on Handbook page 2234 together with suggested tolerances.

Keys Proportioned According to Transmitted Torque.—As previously stated, if key sizes are based on shaft diameter there will be cases where the dimensions of the key will be excessive. This is usually true whenever a gear or pulley transmits only a portion of the total torque capacity of the shaft to which it is keyed. If excessively large keys are to be avoided, it may be advantageous in such cases to base the determination on the torque to be transmitted rather than on the shaft diameter, and to use the dimensions thus determined as a guide in selecting a standard size key.

A key proportioned to transmit a specified torque may fail in service either by shearing or by crushing, depending on the proportions of the key and the manner in which it is fitted to the shaft and hub. The best proportions for a key are those which make it equally resistant to failure by shearing and by crushing. The safe torque in inch-pounds that a key will transmit, based on the allowable shearing stress of the key material, may be found from the formula

$$T_s = L \times W \times \frac{D}{2} \times S_s \qquad (1)$$

The safe torque based on the allowable compressive stress of the key material is found from the formula

$$T_c = L \times \frac{H}{2} \times \frac{D}{2} \times S_c \qquad (2)$$

(For Woodruff keys the amount that the key projects above the shaft is substituted for $H/2$.)

In these formulas, T_s = safe torque in shear; T_c = safe torque in compression; S_s = allowable shearing stress; S_c = allowable compressive stress; L = key length in inches; W = key width in inches; H = key thickness in inches; and D = shaft diameter in inches.

To satisfy the condition that the key be equally resistant to shearing and crushing, T_s should equal T_c. Thus, by equating Formulas (1) and (2), it is found that the width of the keyway in terms of the height of the keyway is

$$W = HS_c \div 2S_s \qquad (3)$$

For the type of steel commonly used in making keys, the allowable compressive stress S_c may be taken as twice the allowable shearing stress S_s of the material if the key is properly fitted on all four sides. By substituting $S_c = 2S_s$ in Formula (3) it will be found that $W = H$, so that for equal strength in compression and shear a square key should be used.

If a rectangular key is used, and the thickness H is less than the width W, then the key will be weaker in compression than in shear so that it is sufficient to check the torque capacity of the key using Formula (2).

Example 6:—A 3-inch shaft is to deliver 100 horsepower at 200 revolutions per minute through a gear keyed to the shaft. If the hub of the gear is 4 inches long, what size key, equally strong in shear and compression, should be used? The allowable compressive stress in the shaft is not to exceed 16,000 pounds per square inch and the key material has an allowable compressive stress of 20,000 pounds per square inch and an allowable shearing stress of 15,000 pounds per square inch.

The first step is to decide on the length of the key. Since the hub of the gear is 4 inches long, a key of the same length may be used. The next step is to determine the torque that the key will have to transmit. Using the formula on Handbook page 302,

$$T = \frac{63,000P}{N} = \frac{63,000 \times 100}{2000} = 31,500 \text{ inch-pounds}$$

To determine the width of the key, based on the allowable shearing stress of the key material, Formula (1) is used.

$$T_s = L \times W \times \frac{D}{2} \times S_s$$

$$31,500 = 4 \times W \times \frac{3}{2} \times 15,000$$

or,

$$W = \frac{31,500 \times 2}{15,000 \times 4 \times 3} = 0.350, \text{ say, } \frac{3}{8} \text{ inch}$$

To determine the thickness of the key, Formula (2) is used. In using this formula, however, it should be noted that if the shaft material has a different allowable compressive stress than the key material, the lower of the two values should be used. In this case, the shaft material has the lower allowable compressive stress and the keyway in the shaft would fail by crushing before the key would. Therefore,

$$T_c = L \times \frac{H}{2} \times \frac{D}{2} \times S_c$$

$$31,500 = 4 \times \frac{H}{2} \times \frac{3}{2} \times 16,000$$

or,

$$H = \frac{31,500 \times 2 \times 2}{4 \times 3 \times 16,000} = 0.656 = \frac{21}{32} \text{ inch}$$

Therefore, the dimensions of the key for equal resistance to failure by shearing and crushing are ⅜ inch wide, ²¹/₃₂ thick, and 4 inches long. If, for some reason, it is desirable to use a key shorter than 4 inches, say 2 inches, then it will be necessary to increase both the width and thickness by a factor of 4 ÷ 2 if equal resistance to shearing and crushing is to be maintained. Thus the width would be ⅜ × ½ = ¾ inch and the thickness would be ²¹/₃₂ × ½ = 1⁵/₁₆ inch for a 2-inch long key.

Set-screws Used to Transmit Torque.—For certain applications it is common practice to use set-screws to transmit torque because they are relatively inexpensive to install and permit axial adjustment of the member mounted on the shaft. However, since set-screws depend primarily on friction and the shearing force at the point of the screw, they are not especially well-suited for high torques or where sudden load changes take place.

One rule for determining the proper size of set-screw states that the diameter of the screw should equal ⁵/₁₆ inch plus one-eighth the shaft diameter. The holding power of set-screws selected by this rule can then be checked using the formula on page 1406 of the Handbook.

PRACTICE EXERCISES FOR SECTION 16

For answers to all practice exercise problems or questions
see Section 22

1. What is the polar section modulus of a shaft 2 inches in diameter?

2. If a 3-inch shaft is subjected to a torsional or twisting moment of 32,800 pound-inches, what is the equivalent torsional or shearing stress?

3. Is the shaft referred to in Exercise 2 subjected to an excessive torsional stress?

4. If a 10-horsepower motor operating at its rated capacity connects by a belt with a 16-inch pulley on the driving shaft of a machine, what is the load tangential to the pulley rim and the resulting twisting moment on the shaft, assuming that the rim speed of the driven pulley is 600 feet per minute?

5. How is the maximum distance between bearings for steel line shafting determined?

6. What are "gib-head" keys and why are they used on some classes of work?

7. What is the distinctive feature of Woodruff keys?

8. What are the advantages of Woodruff keys?

9. If a keyseat ⅜-inch wide is to be milled into a shaft 1½-inch diameter, and if the keyseat depth is ³⁄₁₆ inch (as measured at one side) what is the depth from the top surface of the shaft or the amount to sink the cutter after it grazes the top of the shaft?

SECTION 17

SPLINES

This section of the Handbook shows how to calculate the dimensions of involute splines and how to provide specifications for manufacturing drawings. Many types of mechanical connections between shafts and hubs are available for both fixed and sliding applications. Among these are the ordinary key and keyway (Handbook pages 2234–2257), multiple keys and keyways, three- and four-lobed polygon shaft and hub connections, and involute splines of both inch dimension and metric module sizes.

The major advantages of involute splines are that they may be manufactured on the same equipment used to manufacture gears; they may be used for fixed and interference fit connections as well as for sliding connections; and they are stronger than most other connections with the exception of polygon-shaped members.

The section in the Handbook on involute splines, pages 2017 to 2031, provides tables, data, formulas, and diagrams for American Standard splines made to both inch and metric module systems. Both systems share common definitions of terms, although the symbols used to identify dimensions and angles may differ, as shown on Handbook page 2040. The two systems do not provide for interchangeability of parts; the new metric module standard is the American National Standards Institute version of the International Standards Organization involute spline standard which is based upon metric, not inch, dimensions.

Example 1:—A metric module involute spline pair is required to meet the following specification: pressure angle $\alpha_D = 30°$; module $m = 5$; number of teeth $Z = 32$; fit class = H/h; tolerance class 5 for both the internal and external splines; flat root design for both members: length of engagement of the splines is 100 mm.

Table 2 beginning on Handbook page 2041 provides all the formulas necessary to calculate the dimensions of these splines.

141

Pitch diameter

$$D = mZ = 5 \times 32 = 160 \text{ mm} \tag{1}$$

Base diameter

$$DB = mZ \cos\alpha_D = 160 \times \cos\alpha_D = 160 \times \cos30°$$
$$= 160 \times 0.86603 = 138.5641 \text{ mm} \tag{2}$$

Circular pitch

$$p = \pi m = 3.1416 \times 5 = 15.708 \tag{3}$$

Base pitch

$$p_b = \pi m \cos\alpha_D = \pi \times 5 \times 0.86603 = 13.60350 \tag{4}$$

Tooth thickness modification

$$es = 0 \tag{5}$$

in accordance with the footnote to Table 3, Handbook page 2043 and the Fit Classes paragraph on page 2040 that refers to H/h fits.

Minimum major diameter, internal spline,

$$DEI \text{ min} = m(Z + 1.8) = 5 \times (32 + 1.8) = 169.0000 \tag{6}$$

Maximum major diameter, internal spline,

$$DEI \text{ max} = DEI \text{ min} + (T + \lambda)/ \tan\alpha_D = 169.000 + 0.248/\tan 30°$$
$$= 169.4295 \text{ mm} \tag{7}$$

In this last calculation the value of $(T + \lambda) = 0.248$ for class 7 was calculated using the formula in Table 4, Handbook page 2043, as follows:

$$i^* = 0.001(0.45\sqrt[3]{D} + 0.001D)$$
$$= 0.001(0.45\sqrt[3]{160} + 0.001 \times 160) = 0.00260 \tag{8a}$$

$$i^{**} = 0.001(0.45\sqrt[3]{7.85398} + 0.001 \times 7.85398) = 0.00090 \tag{8b}$$

In this calculation 7.85398 is the value of S_{bsc} calculated from the formula $S_{bsc} = 0.5\pi m$ given on Handbook page 2042.

$$(T + \lambda) = 40i^* + 160 \, i^{**} = 40 \times 0.00260 + 160 \times 000090$$
$$= 0.248 \text{ mm} \tag{8c}$$

Form diameter, internal spline,

$$DFI = m(Z + 1) + 2c_F = 5(32 + 1) + 2 \times 0.1m$$
$$= 5(32 + 1) + 2 \times 0.1 \times 5 = 166 \text{ mm} \tag{9}$$

In the above calculation the value of $c_F = 0.1\ m$ is taken from the diagram on page 2045 or from the formula for form clearance on Handbook page 2042.

Minimum minor diameter, internal spline,

$$DII\ \text{min} = DFE + 2c_F = 154.3502 + 2 \times 0.1 \times 5$$
$$= 155.3502\ \text{mm} \qquad (10)$$

The DFE value of 154.3502 used in this calculation was calculated from the last formula on Handbook page 2041 as follows: $DB = 138.564$ from step (2); $D = 160$ from step (1); $h_s = 0.6\ m = 3.0$ from the last formula on page 2042; $es = 0$ from step (5); $\sin 30° = 0.50000$; $\tan 30° = 0.57735$. Therefore,

$$DFE = 2 \times \sqrt{(0.5 \times 138.564)^2 + \left[0.5 \times 160 \times 0.50000 - \frac{0.6 \times 5 + \left(\dfrac{0.5 \times 0}{0.57735}\right)}{0.50000}\right]^2}$$

$$= 154.3502\ \text{mm} \qquad (11)$$

Maximum minor diameter, internal spline,

$$DII\ \text{max} = DII\ \text{min} + (0.2m^{0.667} - 0.1m^{-0.5}) = 155.3502 + 0.58$$
$$= 155.9302\ \text{mm} \qquad (12)$$

The value 0.58 used in this calculation comes from the footnote (*) to the table on Handbook page 2042.

Circular space width, basic,

$$E_{\text{bsc}} = 0.5\pi m = 0.5 \times 3.1416 \times 5 = 7.854\ \text{mm} \qquad (13)$$

Circular space width, minimum effective,

$$EV\ \text{min} = E_{\text{bsc}} = 7.854\ \text{mm} \qquad (14)$$

Circular space width, maximum actual,

$$E\ \text{max} = EV\ \text{min} + (T + \lambda) = 7.854 + 0.0992\ \text{from step (16c)}$$
$$= 7.9532\ \text{mm} \qquad (15)$$

The value of $(T + \lambda)$ calculated in step (16c) is based upon class 5 fit stated at the beginning of the example. The value calculated in step (8c), on the other hand, is based upon class 7 fit as required by the formula in step (7). For class 5 fit, using the formulas given in Table 4, Handbook page 2043:

$$i^* \quad = 0.00260 \text{ from step (8a)} \tag{16a}$$

$$i^{**} \quad = 0.00090 \text{ from step (8b)} \tag{16b}$$

$$
\begin{aligned}
(T + \lambda) &= 16i^* + 64i^{**} = 16 \times 0.00260 + 64 \times 0.00090 \\
&= 0.0992 \text{ mm}
\end{aligned}
\tag{16c}
$$

Circular space width, minimum actual,

$$E \text{ min} = EV \text{ min} + \lambda = 7.854 + 0.045 = 7.899 \text{ mm} \tag{17}$$

The value of λ used in this formula was calculated from the formulas for class 5 fit in Table 5 and the formula in the text on Handbook page 2044 as follows:

$$F_p = 0.001\,(3.55\sqrt{5 \times 32 \times 3.1416/2} + 9) = 0.065 \text{ mm} \tag{18a}$$

$$f_f = 0.001\,[2.5 \times 5(1 + 0.0125 \times 32) + 16] = 0.034 \text{ mm} \tag{18b}$$

$$F_\beta = 0.001(1 \times \sqrt{100} + 5) = 0.015 \text{ mm} \tag{18c}$$

$$\lambda = 0.6\sqrt{(0.065)^2 + (0.034)^2 + (0.015)^2} = 0.045 \text{ mm} \tag{18d}$$

Circular space width, maximum effective,

$$
\begin{aligned}
EV \text{ max} &= E \text{ max} - \lambda = 7.9532 \text{ from step (15)} - 0.045 \text{ from step} \\
&\text{(18d)} = 7.9082 \text{ mm}
\end{aligned}
\tag{19}
$$

Maximum major diameter, external spline,

$$
\begin{aligned}
DEE \text{ max} &= m(Z + 1) - es/\tan\alpha_D = 5(32 + 1) - 0 \\
&= 165 \text{ mm}
\end{aligned}
\tag{20}
$$

The value 0 in this last calculation is from Table 6, Handbook page 2044 for h class fit.

Minimum major diameter, external spline,

$$DEE \text{ min} = DEE \text{ max} - (0.2m^{0.667} - 0.01m^{-0.5}) = 165 \text{ from step}$$
$$(20) - 0.58 \text{ from (*) on Handbook page 2042} = 164.42 \text{ mm} \tag{21}$$

Maximum minor diameter, external spline,

$$
\begin{aligned}
DIE \text{ max} &= m(Z - 1.8) - es \tan\alpha_D = 5(32 - 1.8) - 0 \\
&= 151 \text{ mm}
\end{aligned}
\tag{22}
$$

The value 0 in this calculation is from Table 6, Handbook page 2044 for h class fit.

Minimum minor diameter, external spline,

DIE min = DIE max $-$ $(T + \lambda)/\tan\alpha_D$ = 151 from step (22)
        $-$ 0.248/tan 30° from step (7) = 151 $-$ 0.4295
        = 150.570 mm $\hspace{4em}$ (23)

Circular tooth thickness, basic,

$$S_{bsc} = 7.854 \text{ mm from step (13)} \hspace{3em} (24)$$

Circular tooth thickness, maximum effective,

SV max = S_{bsc} $-$ es = 7.854 from step (13) $-$ 0 from step (5)
        = 7.845 mm $\hspace{4em}$ (25)

Circular tooth thickness, minimum actual,

S min = SV max $-$ $(T + \lambda)$ = 7.854 from step (25) $-$ 0.0992 from
step (16c) = 7.7548 mm $\hspace{4em}$ (26)

Circular tooth thickness, maximum actual,

S max = SV max $-$ λ = 7.854 from step (25) $-$ 0.045 from step
(18d) = 7.809 mm $\hspace{4em}$ (27)

Circular tooth thickness, minimum effective,

SV min = S min $+$ λ = 7.754 from step (26) $+$ 0.045 from step
(18d) = 7.799 mm $\hspace{4em}$ (28)

Example 2:—As explained on Handbook page 2037, spline gages are used for routine inspection of production parts. However, as part of an analytical procedure to evaluate effective space width or effective tooth thickness, measurements with pins are often used. Measurements with pins are also used for the purpose of checking the actual space width and tooth thickness of splines during the machining process. Such measurements help in making the necessary size adjustments both during the set-up process and as manufacturing proceeds. For the splines calculated in Example 1, what are the pin measurements for the actual tooth thickness and actual space width?

The maximum actual space width for the internal spline is 7.953 mm from step (15) in Example 1. The minimum actual tooth thickness for the external spline is 7.755/mm from step (26).

On Handbook page 2038 there is given a method for calculating pin measurements for splines. This procedure was developed for inch-dimension splines. It may, however, be used for metric module splines simply by replacing P wherever it appears in a formula by

1/m; and by using millimeters instead of inches as dimensional units throughout.

For two-pin measurement *between* pins for the *internal* spline formulas 1, 2, and 3 on Handbook page 2038 are used as follows:

$$\text{inv } \phi_i = 7.953/160 + \text{inv } 30° - 8.64/138.564$$
$$= 0.049706 + 0.053751 - 0.062354 = 0.041103 \quad (1)$$

The numbers used in this calculation are taken from the results in Example 1 except for the involute of 30° which is from the table on page 112 of the Handbook and 8.64 is the diameter of the wire as calculated from the formula on Handbook page 2038, 1.7280/P in which 1/m has been substituted for P to give 1.7280m = 1.7280 × 5 = 8.64. Note that the symbols on page 2038 are not the same as those used in Example 1. This is because the metric standard for involute splines uses different symbols for the same dimensions. The table on page 2040 of the Handbook shows how these different symbols compare.

The value of inv ϕ_i = 0.041103 is used to enter the table on Handbook page 109 to find, by interpolation,

$$\phi_i = 27° \; 36' \; 20'' \quad (2)$$

From the same table find

$$\sec 27° \; 36' \; 20'' = 1.1285 \quad (3)$$

Calculate the measurement between wires:

$$M_i = D_b \sec \phi_i - d_i = 138.564 \times 1.1285 - 8.64$$
$$= 147.729 \text{ mm} \quad (4)$$

For two-pin measurement *over* the teeth of *external* splines, steps 1a, 2a, and 3a on Handbook page 2038 are used as follows:

$$\text{inv } \phi_e = 7.755/160 + 0.053751 + 9.6/138.564 - 3.1416/32$$
$$= 0.073327 \quad (5)$$

Therefore, from Handbook page 114, ϕ_e = 32° 59' and sec 32° 59' = 1.1921.

From formula 3a on Handbook page 2038:

$$M_e = 138.564 \times 1.1921 + 9.6 = 174.782 \text{ mm} \quad (6)$$

The pin diameter 9.6 in this calculation was calculated from the formula in 3a on Handbook page 2038 by substituting 1/m for P in the formula d_e = 1.9200/P = 1.9200m.

Specifying Spline Data on Drawings.—As stated on Handbook page 2029, if the data specified on a spline drawing are suitably arranged and presented in a consistent manner, it is usually not necessary to provide a graphic illustration of the spline teeth. Table 6 on Handbook page 2030 illustrates a flat root spline similar to the one in Example 1 except that it is an inch-dimension spline. The method of presenting drawing data for metric module splines differs somewhat from that shown on page 2030 in that the number of decimal places used for metric spline data is, in some cases, less than that for the corresponding inch-dimension system.

Example 3:—How much of the data calculated or given in Examples 1 and 2 should be presented on the spline drawing?

For the internal spline the data required to manufacture the spline should be presented as follows, including the number of decimal places shown:

INTERNAL INVOLUTE SPLINE DATA

Flat Root Side Fit	Tolerance Class 5H
Number of Teeth	32
Module	5
Pressure Angle	30 deg
Base Diameter	138.5641 REF
Pitch Diameter	160.0000 REF
Major Diameter	169.42 Max
Form Diameter	166.00
Minor Diameter	155.35/155.93
Circular Space Width	
Max Actual	7.953
Min Effective	7.854
Max Measurement Between Pins	147.729 REF
Pin Diameter	8.640

For the external spline:

EXTERNAL INVOLUTE SPLINE DATA

Flat Root Side Fit	Tolerance Class 5h
Number of Teeth	32
Module	5

Pressure Angle	30 deg
Base Diameter	138.5641 REF
Pitch Diameter	160.0000 REF
Major Diameter	164.42/165.00
Form Diameter	154.35
Minor Diameter	150.57 MIN
Circular Tooth Thickness	
Max Effective	7.854
Min Actual	7.809
Min Measurement Over Pins	174.782 REF
Pin Diameter	9.6

PRACTICE EXERCISES FOR SECTION 17

For answers to all practice exercise problems or questions
see Section 22

1. What is the difference between a "soft" conversion of a standard and a "hard" system?

2. The standard for metric module splines does not include a major diameter fit. What standard does provide for a major diameter fit?

3. What is an involute serration and is it still so called in American standards?

4. What are some of the advantages of involute splines?

5. What is the meaning of the term "effective tooth thickness"?

6. What advantage is there in using an odd number of spline teeth?

7. If a spline connection is misaligned, fretting can occur at certain combinations of torque, speed, and misalignment angle. Is there any method for diminishing such damage?

8. For a given design of spline is there a method for estimating the torque capacity based upon wear? Shearing stress?

9. What does REF following a dimension of a spline mean?

10. Why are fillet root splines preferred in some cases over flat root splines?

SECTION 18

PROBLEMS IN DESIGNING AND CUTTING GEARS

HANDBOOK Pages 1765–1942

In the design of gearing, there may be three distinct types of problems. These are: (1) determining the relative sizes of two or more gears to obtain a given speed or series of speeds; (2) determining the pitch of the gear teeth so that they will be strong enough to transmit a given amount of power; (3) calculating the dimensions of a gear of a given pitch, such as the outside diameter, the depth of the teeth, and other dimensions needed in cutting the gear.

When the term diameter is applied to a spur gear, the pitch diameter is generally referred to and not the outside diameter. In calculating the speeds of gearing, the pitch diameters are used and not the outside diameters, because when gears are in mesh the imaginary pitch circles roll in contact with each other.

Calculating Gear Speeds.—The simple rules for calculating the speeds of pulleys beginning on Handbook page 2269 may be applied to gearing, provided either the pitch diameters of the gears or the numbers of teeth are substituted for the pulley diameters. Information on gear speeds, especially as applied to compound trains of gearing, also will be found in the section dealing with lathe change gears beginning on Handbook page 1672. When gear trains must be designed to secure unusual or fractional gear ratios, the tables of logarithms of gear ratios—Handbook pages 1679 to 1702—will be found very useful. A practical application of these tables is shown by a number of examples beginning on Handbook page 1676.

Planetary or epicyclic gearing is an increasingly important class of power transmission in various industries because of compactness, efficiency, and versatility. The rules for calculating rotational speeds and ratios are different from those for other types of gearing. Formulas for the most commonly used types of planetary gears are provided on Handbook pages 1937 to 1940.

Example 1:—The following example illustrates the method of calculating the speed of a driven shaft in a combination belt and gear

149

drive when the diameters of the pulleys and the pitch diameters of
the gears are known, and the number of revolutions per minute of
the driving shaft is given. If driving pulley *A*, Fig. 1, is 16 inches in
diameter, and driven pulley *B*, 6 inches in diameter, and the pitch
diameter of driving gear *C* is 12 inches, driving gear *D* is 14 inches,
driven gear *E*, 7 inches, driven gear *F*, 6 inches, and driving pulley
A makes 60 revolutions per minute, determine the number of revolutions per minute of *F*.

$$\frac{16 \times 12 \times 14}{6 \times 7 \times 6} \times 60 = 640 \text{ revolutions per minute}$$

The calculations required in solving problems of this kind can be
simplified if the gears are considered as pulleys having diameters
equal to their pitch diameters. When this is done, the rules that apply
to compound belt drives can be used in determining the speed or
size of the gears or pulleys.

If the number of teeth in each gear is substituted for its pitch
diameter, the result will be the same as when the pitch diameters
are used.

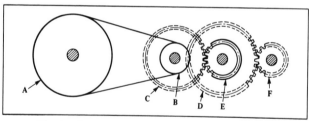

Fig. 1.   Combination Pulley and Compound Gear Drive

Example 2:—If driving spur gear *A* (Fig. 2) makes 336 revolutions
per minute and has 42 teeth, driven spur gear *B*, 21 teeth, driving
bevel gear *C*, 33 teeth, driven bevel gear *D*, 24 teeth, driving worm
E, one thread, and driven worm-wheel *F*, 42 teeth, determine the
number of revolutions per minute of *F*.

When a combination of spur, bevel, and worm gearing is employed
to transmit motion and power from one shaft to another, the speed
of the driven shaft can be found by the following method: Consider
the worm as a gear having one tooth if it is single-threaded and as

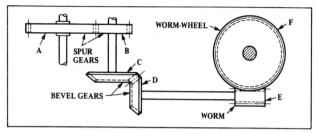

Fig. 2. Combination of Spur, Bevel, and Worm Gearing

a gear having two teeth if double-threaded, etc. When this is done, the speed of the driving shaft can be found by applying the rules for ordinary compound spur gearing. In this example,

$$\frac{42 \times 33 \times 1}{21 \times 24 \times 42} \times 336 = 22 \text{ revolutions per minute}$$

If the pitch diameters of the gears are used instead of the number of teeth in making calculations, the worm should be considered as a gear having a pitch diameter of 1 inch, if single-threaded, and 2 inches if a double-threaded worm, etc.

Example 3:—If a worm is triple-threaded and makes 180 revolutions per minute, and the worm-wheel is required to make 5 revolutions per minute, determine the number of teeth in the worm-wheel.

Rule: Multiply the number of threads in the worm by its number of revolutions per minute, and divide the product by the number of revolutions per minute of the worm-wheel. Applying this rule,

$$\frac{3 \times 180}{5} = 108 \text{ teeth}$$

Example 4:—A 6-inch grinding machine with a spindle speed of 1773 revolutions per minute, for a recommended peripheral speed of 6500 feet per minute (as figured for a full-size 14-inch wheel for this size of machine), has two steps on the spindle pulley; the large step is 5.5 inches in diameter and the small step, 4 inches. What should be the minimum diameter of the wheel before the belt is shifted to the smaller step in order to obtain again a peripheral wheel speed of 6500 feet per minute?

As the spindle makes 1773 revolutions per minute when the belt is on the large pulley, its speed with the belt on the smaller may be determined as follows: 5.5 : 4 = x : 1773, or (5.5 × 1773)/4 = 2438 revolutions per minute, approximately. To obtain the same peripheral speed as when the belt is on the large pulley, the diameters of the grinding wheel should be 14 : x = 2438 : 1773, or (14 × 1773)/2438 = 10.18 inches. Therefore, when the grinding wheel has been worn down to a diameter of 10.18 inches, or approximately 10³⁄₁₆ inches, the spindle belt should be shifted to the smaller step of the spindle pulley to obtain a peripheral speed of 6500 feet per minute. The method used in this example may be reduced to a formula for use with any make of grinding machine having a two-step spindle pulley.

Let D = diameter of wheel, full size;

D_1 = diameter of wheel, reduced size;

d = diameter of large pulley step;

d_1 = diameter of small pulley step;

V = spindle R.P.M., using large pulley step;

v = spindle R.P.M., using small pulley step.

Then
$$v = \frac{dV}{d_1}; \qquad D_1 = \frac{DV}{v}$$

Example 5:—Planetary gear sets are widely used in power transmission because of their compactness and relatively high efficiency when properly designed. The simple planetary configuration shown in Fig. 10 on Handbook page 1939 is typical of high-efficiency designs. If $A = 20$ and $C = 40$, what is the rotation of the driver D per revolution of the follower?

Using the formula given on Handbook page 1939,

$$D = 1 + \frac{C}{A} = 1 + \frac{40}{20} = 3$$

Example 6:—If in Example 5 the diameter of the fixed gear is doubled to $C = 80$, what effect does that produce in the rotation of the driver D?

$$D = 1 + \frac{80}{20} = 5$$

Note that doubling the size of the fixed gear C does not double the ratio or the driver speed of the gear set because the overall ratio is always 1 plus the ratio of C to A.

Example 7:—The compound type of planetary gear shown in Fig. 13 on Handbook page 1939 can provide high reduction ratios although the efficiency decreases as the ratio increases. What is the rotation of the follower *F* when $B = 61$, $C = 60$, $x; = 19$, and $y = 20$?

$$F = 1 - \left(\frac{C \times x}{y \times B}\right) = 1 - \left(\frac{60 \times 19}{20 \times 61}\right) = 1 - \frac{57}{61} = 0.06557$$

Example 8:—In Example 7, what is the rotation of the driver per revolution of the follower?

$$\text{Driver} = \frac{1}{\text{Follower}} = \frac{1}{0.06557} = 15.25$$

Note that in compound planetary gear drives the sum of meshing tooth pairs must be equal for proper meshing. Thus, $C + y = x + B$.

Diametral Pitch of a Gear.—The diametral pitch represents the number of gear teeth for each inch of pitch diameter and, therefore, equals the number of teeth divided by the pitch diameter. The term diametral pitch as applied to bevel gears has the same meaning as in the case of spur gears. This method of basing the pitch on the relation between the number of teeth and the pitch diameter, is used almost exclusively in connection with cut gearing and to some extent for cast gearing. The circular pitch or the distance between the centers of adjacent teeth measured along the pitch circle is used for cast gearing, but very little for cut gearing excepting very large sizes. If 3.1416 is divided by the diametral pitch, the quotient equals the circular pitch, or if the circular pitch is known, the diametral pitch may be found by dividing 3.1416 by the circular pitch.

The pitch of gear teeth may depend primarily upon the strength required to transmit a given amount of power.

Power Transmitting Capacity of Bevel Gears.—As pointed out in the text on page 1854 of the Handbook, the design of a pair of bevel gears to meet a given set of operating conditions is best accomplished in four steps: (1) Determine the design load upon which the sizes of the bevel gears will be based; (2) by means of the series of design charts given in the Handbook, select approximate gear and pinion sizes to satisfy the design load requirements; (3) check the surface durability of the gears selected in the second step by using the surface durability formula on Handbook page 1857; and (4) check the strength of the gears by using the formula given on Handbook page 1863. In cases where it is desired to check the capacity of an existing bevel gear drive, only steps (3) and (4) are necessary.

Dimensions and Angles Required in Producing Gears.—Many of the rules and formulas given in the gear section of the Handbook beginning on page 1765 are used in determining tooth dimensions, gear blank sizes, also angles in bevel, helical, and worm gearing. These dimensions or angles are required on the working drawings used in connection with machining operations, such as turning gear blanks and cutting the teeth.

Example 9:—If a spur gear is to have 40 teeth of 8 diametral pitch, to what diameter should the blank be turned? Applying Formula 7a, Handbook page 1771, (40 + 2)/8 = 5.25 inches. Therefore, the outside diameter of this gear or the diameter to which the blank would be turned is 5¼ inches.

In the case of internal spur gears, the inside diameter to which the gear blank would be bored may be obtained by subtracting 2 from the number of teeth, and dividing the remainder by the diametral pitch.

Example 10:—A sample spur gear has 22 teeth, and the outside diameter, or diameter measured across the tops of the teeth, is 6 inches. Determine the diametral pitch.

According to Formula No. 7a, Handbook page 1771,

$$D_O = \frac{N + 2}{P}; \text{ hence,}$$

$$P = \frac{N + 2}{D_O} = \frac{22 + 2}{6} = 4 \text{ diametral pitch}$$

The table, Handbook page 1771, also shows that when the sample gear has American Standard Stub teeth, Formula 8a should be used to determine the outside diameter, or diametral pitch.

Example 11:—A 25-degree involute full-depth spur gear is to be produced by hobbing. How is the hob tip radius found?

As shown on Handbook page 1805, the maximum hob tip radius, r_c (max.), is found by the formula:

$$r_c \text{ (max.)} = \frac{0.785398 \cos \phi - b \sin \phi}{1 - \sin \phi}$$

where ϕ is the pressure angle, in this case, 25°, and b is the dedendum constant which is 1.250 according to Table 2 on Handbook page 1772. Thus,

$$r_c \text{ (max.)} = \frac{0.785398 \times 0.90631 - 1.25 \times 0.42262}{1 - 0.42262}$$

$$= 0.3179 \text{ inch for a 1 diametral pitch gear}$$

Example 12:—If a 20-degree involute full-depth pinion having 24 teeth of 6 diametral pitch is to mesh with a rack, determine the whole depth of the rack teeth and the linear pitch of the teeth.

The teeth of a rack are of the same proportions as the teeth of a spur gear or pinion which is intended to mesh with the rack; hence the pitch of the rack teeth is equal to the circular pitch of the pinion, and is found by dividing 3.1416 by the diametral pitch.

The pitch = 3.1416 ÷ 6 = 0.5236 inch = linear pitch of rack for meshing with a pinion of 6 diametral pitch. This dimension (0.5236) represents the distance that the cutter would be indexed when milling rack teeth, or distance that the planer tool would be moved for cutting successive teeth in case the planer were used. The whole depth of a full-depth rack tooth of 20-degree pressure angle equals 2.157 divided by the diametral pitch of the meshing gear, or the whole depth equals the circular pitch multiplied by 0.6866. In this case, the circular pitch is 0.5236 and the whole depth equals 0.5236 × 0.6866 = 0.3595 inch.

Example 13:—If the teeth of a spur gear are to be cut to a certain diametral pitch, is it possible to obtain any diameter that may be desired? Thus, if the diametral pitch is 4, is it possible to make the pitch diameter 5⅛ inches?

The diametral pitch system is so arranged as to provide a series of tooth sizes, just as the pitches of screw threads are standardized. Inasmuch as there must be a whole number of teeth in each gear, it is apparent that gears of a given pitch vary in diameter according to the number of teeth. Suppose, for example, that a series of gears are of 4 diametral pitch. Then the pitch diameter of a gear having, say, 20 teeth will be 5 inches; 21 teeth, 5¼ inches; 22 teeth, 5½ inches, and so on, It will be seen that the increase in diameter for each additional tooth is equal to ¼ inch for 4 diametral pitch. Similarly for 2 diametral pitch the variations for successive numbers of teeth would equal ½ inch, and for 10 diametral pitch the variations would equal ¹⁄₁₀ inch, etc.

The center-to-center distance between two gears is equal to one-half the total number of teeth in the gears divided by the diametral pitch. While it may be desirable at times to have a center distance which cannot be obtained exactly by any combination of gearing of given diametral pitch, this is an unusual condition and ordinarily the designer of a machine can alter the center distance whatever slight amount may be required for gearing of the desired ratio and pitch. By using a standard system of pitches all calculations are simplified, and it is also possible to obtain the benefits of standardization in the manufacturing of gears and gear-cutters.

Proportioning Spur Gears when Center Distance is Fixed.—If the center-to-center distance between two shafts is fixed and it is desired to use gears of a certain pitch, the number of teeth in each gear for a given speed may be determined as follows: Since the gears must be of a certain pitch, the total number of teeth available should be determined and then the number of teeth in the driving and the driven gears. The total number of teeth equals twice the product of the center distance multiplied by the diametral pitch. If the center distance is 6 inches and the diametral pitch 10, the total number of teeth equals $6 \times 2 \times 10 = 120$ teeth. The next step is to find the number of teeth in the driving and the driven gears for a given rate of speed.

Rule: Divide the speed of the driving gear in revolutions per minute by the speed of the driven gear and add one to the quotient. Next divide the total number of teeth in both gears by the sum previously obtained, and the quotient will equal the number of teeth in the driving gear. This number subtracted from the total number of teeth will equal the number of teeth required in the driven gear.

Example 14:—If the center-to-center distance is 6 inches and the diametral pitch is 10, the total number of teeth available will be 120. If the speeds of the driving and the driven gears are to be 100 and 60 revolutions per minute, respectively, find the number of teeth for each gear.

$$^{100}\!/_{60} = 1\tfrac{2}{3} \text{ and } 1\tfrac{2}{3} + 1 = 2\tfrac{2}{3}$$

$$120 \div 2\tfrac{2}{3} = {}^{120}\!/_1 \times \tfrac{3}{8} = 45 = \text{number of teeth in driving gear}$$

The number of teeth in the driven gear equals $120 - 45 = 75$ teeth.

When the center distance and velocity ratios are fixed by some essential construction of a machine, it is often impossible to use standard diametral pitch gear teeth. If cast gears are to be used, it does not matter so much, as a patternmaker can lay out the teeth according to the pitch desired, but if cut gears are required, an effort should be made to alter the center distance so that standard diametral pitch cutters can be used since these are usually carried in stock.

Dimensions of Generated Bevel Gears.—*Example 15:*—Find all of the dimensions and angles necessary to manufacture a pair of straight bevel gears if the number of teeth in the pinion is 16, the number of teeth in the mating gear is 49, the diametral pitch is 5, and the face width is 1.5 inches. The gears have a 20-degree pressure angle, a 90-degree shaft angle, and are to be in accordance with the Gleason System.

On Handbook page 1845 is a table of formulas for Gleason System 20-degree pressure angle straight bevel gears with 90-degree shaft angle. These formulas are given in the same order as is normally used in computation. Computations of the gear dimensions should be arranged as follows for neatness and to establish a consistent procedure when calculations for bevel gears are required frequently.

Number of pinion teeth = 16   (1)
Number of gear teeth   = 49   (2)
Diametral pitch   = 5   (3)
Face width   = 1.5   (4)
Pressure angle   = 20°   (5)
Shaft angle   = 90°   (6)

	PINION	GEAR	
Working depth	$\dfrac{2.000}{5} = 0.400$	Same as pinion	(7)
Whole depth	$\dfrac{2.188}{5} + 0.002 = 0.440$	Same as pinion	(8)
Pitch diameter	$^{16}\!/_5 = 3.2000$	$^{49}\!/_5 = 9.8000$	(9)
Pitch angle	$\tan^{-1} {}^{16}\!/_{49} = 18° 5'$ *Note:* $\tan^{-1} {}^{16}\!/_{49}$ should be read as "the angle whose tangent is 16 ÷ 49"	$90° - 18° 5' = 71° 55'$	(10)

	PINION	GEAR
Cone distance	$\dfrac{9.8000}{2 \times \sin 71° 55'} =$ 5.1546	Same as pinion $\qquad$ (11)
Circular pitch	$\dfrac{3.1416}{5} = 0.6283$	Same as pinion $\qquad$ (12)
Addendum	$0.400 - 0.116 =$ 0.284	$\dfrac{0.540}{5} + \dfrac{0.460}{5(^{49}/_{16})^2} =$ 0.118 (13)
Dedendum	$\dfrac{2.188}{5} - 0.284 =$ 0.1536	$\dfrac{2.188}{5} - 0.118 =$ 0.3214 (14)
Clearance	$0.440 - 0.400 =$ 0.040	Same as pinion $\qquad$ (15)
Dedendum angle	$\tan^{-1} \dfrac{0.1536}{5.1546} =$ $1° 42'$	$\tan^{-1} \dfrac{0.3214}{5.1546} = 3° 34'$ (16)
Face angle of blank	$18° 5' + 3° 34' =$ $21° 39'$	$71° 55' + 1° 42' =$ $73° 37'$ (17)
Root angle	$18° 5' - 1° 42' =$ $16° 23'$	$71° 55' - 3° 34' =$ $68° 21'$ (18)
Outside diameter	$3.2000 + 2 \times$ $0.284 \cos 18° 5' =$ 3.740	$9.8000 + 2 \times$ $0.118 \cos 71° 55' =$ 9.875 (19)
Pitch apex to crown	$\dfrac{9.8000}{2} -$ $0.284 \sin 18° 5' =$ 4.812	$\dfrac{3.2000}{2} -$ $0.118 \sin 71° 55' =$ 1.488 (20)

	PINION	GEAR
Circular thick-ness	$0.6283 - 0.2530 =$ 0.3753	$\dfrac{0.6283}{2} -$ $\dfrac{(0.284 - 0.118)\tan 20° -}{5}$ $\dfrac{0.038 \text{ (chart 1)}}{5} =$ $\qquad 0.2461 \quad (21)$
Backlash		$0.006 \qquad\qquad (22)$
Chordal thick-ness	$0.3753 -$ $\dfrac{(0.3753)^3}{6 \times (3.2000)^2} -$ $\dfrac{0.006}{2} = 0.371$	$0.2461 -$ $\dfrac{(0.2461)^3}{6 \times (9.8000)^2} -$ $\dfrac{0.006}{2} = 0.243 \quad (23)$
Chordal addendum	$0.284 +$ $\dfrac{0.3753^2 \cos 18° 5'}{4 \times 3.2000}$ $= 0.294$	$0.118 +$ $\dfrac{0.2461^2 \cos 71° 55'}{4 \times 9.8000} =$ $\qquad 0.118 \quad (24)$

The tooth angle (Item 25) is a machine setting and is only computed if a Gleason two-tool type straight bevel gear generator is to be used. For a further description of tooth angle, Item 25, see Handbook page 1844.

Dimensions of Milled Bevel Gears.—As explained on Handbook page 1867, the tooth proportions of milled bevel gears differ in some respects from those of generated bevel gears. To take these differences into account a separate table of formulas is given on Handbook page 1868 for use in calculating dimensions of milled bevel gears.

Example 16:—Compute the dimensions and angles of a pair of mating bevel gears that are to be cut on a milling machine using rotary formed milling cutters if the data given is as follows:

$$\text{Number of pinion teeth} = 15$$
$$\text{Number of gear teeth} = 60$$
$$\text{Diametral pitch} = 3$$
$$\text{Pressure angle} = 14\tfrac{1}{2}°$$
$$\text{Shaft angle} = 90°$$

Using the formulas on Handbook page 1868,

$$\tan \alpha_P = 15 \div 60 = 0.25000 = \tan 14° 2.2', \text{ say, } 14° 2'$$
$$\alpha_G = 90° - 14° 2' = 75° 57.8', \text{ say, } 75° 58'$$
$$D_P = 15 \div 3 = 5.000 \text{ inches}$$
$$D_G = 60 \div 3 = 20.000 \text{ inches}$$
$$S = 1 \div 3 = 0.3333 \text{ inch}$$
$$S + A = 1.157 \div 3 = 0.3857 \text{ inch}$$
$$W = 2.157 \div 3 = 0.7190 \text{ inch}$$
$$T = 1.571 \div 3 = 0.5236 \text{ inch}$$
$$C = \frac{5.000}{2 \times 0.24249} = 10.308 \text{ inch}$$

(In determining C, the sine of the unrounded value of α_P, $14° 2.2'$, is used.)

$$F = 8 \div 3 = 2\tfrac{2}{3}, \text{ say, } 2\tfrac{5}{8} \text{ inches}$$
$$s - 0.3333 \times \frac{10.308 - 2\tfrac{5}{8}}{10.308} = 0.2484 \text{ inch}$$
$$t = 0.5236 \times \frac{10.308 - 2\tfrac{5}{8}}{10.308} = 0.3903 \text{ inch}$$
$$\tan \theta = 0.3333 \div 10.308 = \tan 1° 51'$$
$$\tan \phi = 0.3857 \div 10.308 = \tan 2° 9'$$
$$\gamma_P = 14° 2' + 1° 51' = 15° 53'$$
$$\gamma_G = 75° 58' + 1° 51' = 77° 49'$$
$$\delta_P = 90° - 15° 53' = 74° 7'$$
$$\delta_G = 90° - 77° 49' = 12° 11'$$
$$\zeta_P = 14° 2' - 2° 9' = 11° 53'$$
$$\zeta_G = 75° 58' - 2° 9' = 73° 49'$$
$$K_P = 0.3333 \times 0.97015 = 0.3234 \text{ inch}$$
$$K_G = 0.3333 \times 0.24249 = 0.0808 \text{ inch}$$
$$O_P = 5.000 + 2 \times 0.3234 = 5.6468 \text{ inches}$$
$$O_G = 20.000 + 2 \times 0.0808 = 20.1616 \text{ inches}$$
$$J_P = \frac{5.6468}{2} \times 3.5144 = 9.9226 \text{ inches}$$
$$J_G = \frac{20.1616}{2} \times 0.21590 = 2.1764 \text{ inches}$$
$$j_P = 9.9226 \times \frac{10.3097 - 2\tfrac{5}{8}}{10.3097} = 7.3961 \text{ inches}$$
$$j_G = 2.1764 \times \frac{10.3097 - 2\tfrac{5}{8}}{10.3097} = 1.6222 \text{ inches}$$

$$N'_P = \frac{15}{0.97015} = 15.4, \text{ say, 15 teeth}$$

$$N'_G = \frac{60}{0.24249} = 247 \text{ teeth}$$

If these gears are to have uniform clearance at the bottom of the teeth, in accordance with the recommendation given in the last paragraph on Handbook page 1867, then the cutting angles ζ_P and ζ_G should be determined by subtracting the addendum angle from the pitch cone angles. Thus,

$$\zeta_P = 14° \, 2' - 1° \, 51' = 12° \, 11'$$

$$\zeta_G = 75° \, 58' - 1° \, 51' = 74° \, 7'$$

Selection of Formed Cutters for Bevel Gears.—*Example 17:*—In Example 16, the numbers of teeth for which to select the cutters were calculated as 15 and 247 for the pinion and gear respectively. Therefore, as explained on page 1871 of the Handbook, the cutters selected from the table on page 1794 are the No. 7 and the No. 1 cutters. As further noted on page 1871, bevel gear milling cutters may be selected directly from the table beginning on page 1869, when the shaft angle is 90 degrees, instead of using the computed value of N' to enter the table on page 1794. Thus, in this case, for a 15-tooth pinion and a 60-tooth gear, the table on page 1870 shows that the numbers of the cutters to use are 1 and 7 for gear and pinion, respectively.

Pitch of Hob for Helical Gears.—*Example 18:*—A helical gear that is to be used for connecting parallel shafts has 83 teeth, a helix angle of 7 degrees, and a pitch diameter of 47.78 inches. Determine the pitch of hob to use in cutting this gear.

As explained on Handbook page 1907, the normal diametral pitch or the pitch of the hob is determined as follows: The transverse diametral pitch equals $83 \div 47.78 = 1.737$. The cosine of the helix angle of the gear (7 degrees) is 0.99255; hence the normal diametral pitch equals $1.737 \div 0.99255 = 1.75$; therefore, a hob of 1¾ diametral pitch should be used. This hob is the same as would be used for spur gears of 1¾ diametral pitch, and it will cut any spur or helical gear of that pitch regardless of the number of teeth, provided 1¾ is the diametral pitch of the spur gear and the normal diametral pitch of the helical gear.

Determining Contact Ratio.—As pointed out on Handbook page 1804 if a smooth transfer of load is to be obtained from one pair of teeth to the next pair of teeth as two mating gears rotate under load, the contact ratio must be well over 1.0. Usually, this ratio should be 1.4 or more, although in extreme cases it may be as low as 1.15.

Example 19:—Find the contact ratio for a pair of 18-diametral pitch, 20-degree pressure angle gears, one having 36 teeth and the other 90 teeth. From formula (1) given on Handbook page 1803:

$$\cos A = \frac{90 \times \cos 20°}{5.111 \times 18}$$

$$= \frac{90 \times 0.93969}{91.9998}$$

$$\cos A = 0.91926 \text{ and}$$

$$A = 23°11'$$

From formula (4) on Handbook page 1804:

$$\cos a = \frac{36 \times \cos 20°}{2.1111 \times 18}$$

$$= \frac{36 \times 0.93969}{37.9998}$$

$$\cos a = 0.89024 \text{ and}$$

$$a = 27°6'$$

From formula (5) on Handbook page 1804:

$$\tan B = \tan 20° - \frac{36}{90} (\tan 27°6' - \tan 20°)$$

$$= 0.36397 - \frac{36}{90} (0.51172 - 0.36397)$$

$$\tan B = 0.36397 - 0.05910$$

$$= 0.30487$$

From formula (7a) on Handbook page 1804, the contact ratio m_f is found:

$$m_f = \frac{90}{6.28318} (0.42826 - 0.30487)$$

$$= 1.77$$

which is satisfactory.

Dimensions Required when Using Enlarged Fine-pitch Pinions.—
On Handbook pages 1799–1801 there are tables of dimensions for
enlarged fine-pitch pinions. These tables show how much the dimensions of enlarged pinions must differ from standard when the
number of teeth is small, and undercutting of the teeth is to be
avoided.

Example 20:—If a 10- and a 31-tooth mating pinion and gear of
20 diametral pitch and 14½° pressure angle have both been enlarged
to avoid undercutting of the teeth, what increase over standard center
distance is required?

$$\text{The standard center distance} = \frac{n + N}{2P}$$

$$= \frac{10 + 31}{2 \times 20} = 1.0250 \text{ inch}$$

The amount by which the center distance must be increased over
standard can be obtained by taking the sum of the amounts shown
in the eighth column of Table 3b on Handbook page 1799 and dividing
this sum by the diametral pitch. Thus, in this case, the increase over
standard center distance is $\dfrac{0.6866 + 0.0283}{20} = 0.0357$ inch.

Example 21:—At what center distance would the gears in Example
20 have to be meshed if there were to be no backlash?

Obtaining the two thicknesses of both gears at the standard pitch
diameters from Table 3b on Handbook page 1799, dividing them by
20, and using the formulas at the top of Handbook page 1803:

$$\text{inv } \phi_1 = \text{inv } 14\frac{1}{2}° + \frac{20\,(.09630 + .07927) - 3.1416}{10 + 31}$$

The involute of 14½° is found on Handbook page 96 to be
0.0055448, hence:

$$\text{inv } \phi_1 = 0.0055448 + 0.0090195$$

$$= 0.0145643$$

Referring to the table on Handbook page 101:

$$\phi_1 = 19° \ 51' \ 6''$$

$$C = \frac{10 + 31}{2 \times 20}$$

$$= 1.025$$

$$C_1 = \frac{\cos 14\frac{1}{2}°}{\cos 19° \, 51' \, 6''} \times 1.025$$

$$= \frac{0.96815}{0.94057} \times 1.025$$

$$C_1 = 1.0551 \text{ inch}$$

End Thrust of Helical Gears Applied to Parallel Shafts.—*Example 22:*—The diagrams on Handbook pages 1910 to 1912 show the application of helical or spiral gears to parallel shaft drives. If 7 horsepower is to be transmitted at a pitch-line velocity of 200 feet per minute, determine the end thrust in pounds, assuming that the helix angle of the gear is 15 degrees.

In order to determine the end thrust of helical gearing as applied to parallel shafts, first calculate the tangential load on the gear teeth.

$$\text{Tangential load} = \frac{33,000 \times 7}{200} = 1155 \text{ pounds}$$

(This formula is derived from the formulas for power given on Handbook page 175.)

The axial or end thrust may now be determined approximately by multiplying the tangential load by the tangent of the tooth angle. Thus, in this instance, the thrust = 1155 × tan 15 degrees = about 310 pounds. (Note that this formula agrees with the one on Handbook page 146 for determining force P parallel to base of inclined plane.) The end thrust obtained by this calculation will be somewhat greater than the actual end thrust, because frictional losses in the shaft bearings, etc., have not been taken into account, although a test on a helical gear set, with a motor drive, showed that the actual thrust of the 7½-degree helical gears tested, was not much below the values calculated as just explained.

According to most text-books, the maximum angle for single helical gears should be about 20 degrees, although one prominent manufacturer mentions that the maximum angle for industrial drives ordinarily does not exceed 10 degrees, and this will give quiet running without excessive end thrust. On some of the heavier single helical gearing used for street railway transmissions, etc., an angle of 7 degrees is employed.

Dimensions of Wormgear Blank and the Gashing Angle.—*Example 23:*—A wormgear having 45 teeth is to be driven by a double threaded worm having an outside diameter of 2½ inches and a lead of 1 inch, the linear pitch being ½ inch. The throat diameter and throat radius of the wormgear are required, and also the angle for gashing the blank.

The throat diameter is found by applying rule or formula No. 28 on Handbook page 1888. The addendum of the worm thread equals the linear pitch multiplied by 0.3183, and in this case 0.5 × 0.3183 = 0.1591 inch. The pitch diameter of the wormgear = 45 × 0.5 ÷ 3.1416 = 7.162 inches; hence, the throat diameter equals 7.162 + 2 × 0.1591 = 7.48 inches.

The radius of the wormgear throat is found by subtracting twice the addendum of the worm thread from ½ the outside diameter of the worm. The addendum of the worm thread equals 0.1591 inch, and the radius of the throat, therefore, equals (2.5 ÷ 2) − 2 × 0.1591 = 0.931 inch.

When a wormgear is hobbed in a milling machine, gashes are milled before the hobbing operation. The table must be swiveled around while gashing, the amount depending upon the relation between the lead of the worm thread and the pitch circumference. The first step is to find the circumference of the pitch circle of the worm. The pitch diameter equals the outside diameter minus twice the addendum of the worm thread; hence, the pitch diameter equals 2.5 − 2 × 0.1591 = 2.18 inches, and the pitch circumference equals 2.18 × 3.1416 = 6.848 inches.

Next divide the lead of the worm thread by the pitch circumference to obtain the tangent of the desired angle, and then refer to a table of tangents to determine what this angle is. In this case it is 1 ÷ 6.848 = 0.1460 which is the tangent of 8⅓ degrees, nearly. Therefore, the table of the milling machine is set at an angle of 8⅓ degrees from its normal position.

Change Gear Ratio for Diametral-pitch Worms.—*Example 24:* In cutting worms to a given diametral pitch, the ratio of the change gears as given on Handbook page 1899, is $\dfrac{22 \times \text{Threads per Inch}}{7 \times \text{Diametral Pitch}}$.

Explain why the constants 22 and 7 are used.

The reason why the constants 22 and 7 are used in determining the ratio of change-gears for cutting worm threads, is because $^{22}/_{7}$ equals, very nearly, 3.1416, which is the circular pitch equivalent to 1 diametral pitch.

Assume that the diametral pitch of the wormgear is 5, and the lathe screw constant is 4. (See Handbook page 1672 for the meaning of "lathe screw constant.") Then, $\dfrac{4 \times 22}{5 \times 7} = \dfrac{88}{35}$. If this simple combination of gearing were used, the gear on the stud would have 88 teeth, and the gear on the lead screw, 35 teeth. Of course, any other combination of gearing having this same ratio could be used, as for example, the following compound train of gearing: $\dfrac{24 \times 66}{30 \times 21}$.

If the lathe screw constant is 4, as previously assumed, then the number of threads per inch obtained with gearing having a ratio of $\dfrac{88}{35} = \dfrac{4 \times 35}{88} = 1.5909$; hence, the pitch of the worm thread equals $1 \div 1.5909 = 0.6284$ inch, which is the circular pitch equivalent to 5 diametral pitch, correct to within 0.0001 inch.

Bearing Loads Produced by Bevel Gears (Handbook page 1853).— In applications where bevel gears are used, not only must the gears be proportioned with regard to the power to be transmitted, but also the bearings supporting the gear shafts must be of adequate size and design to sustain the radial and thrust loads that will be imposed on them. Assuming that suitable gear and pinion proportions have been selected, the next step is to compute the loads needed to determine whether or not adequate bearings can be provided. To find the loads on the bearings, first, use the formulas on Handbook page 1854 to compute the tangential, axial, and separating components of the load on the tooth surfaces. Secondly, use the principle of moments, together with the components determined in the first step, to find the radial loads on the bearings. To illusrate the procedure, the following example will be used.

Example 25:—A 16-tooth left-hand spiral pinion rotating clockwise at 1800 rpm transmits 71 horsepower to a 49-tooth mating gear. If the pressure angle is 20 degrees, the spiral angle is 35 degrees, the face width is 1.5 inches, and the diametral pitch is 5, what are the radial and thrust loads that govern the selection of bearings?

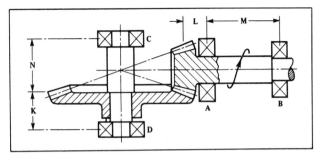

Fig. 3. Diagram Showing Location of Bearings for Bevel Gear Drive in Example 25

In Fig. 3, the locations of the bearings for the gear shafts are shown. It should be noted that distances K, L, M, and N are measured from the center line of the bearings and from the midfaces of the gears at their mean pitch diameters. In this example it will be assumed that these distances are given and are as follows: $K = 2.5$ inches; $N = 3.5$ inches; $L = 1.5$ inches; and $M = 5.0$ inches.

Other quantities that will be required in the solution of this example are the pitch diameter, pitch angle, and mean pitch diameter of both the gear and pinion. These are computed using formulas given in the Handbook.

Using Formula 9 from page 1849,

Pitch diam. of pinion $d = 3.2$ inches
Pitch diam. of gear $D = 9.8$ inches

Using Formula 10 from page 1849,

Pitch angle of pinion $\gamma = 18°\ 5'$
Pitch angle of gear $\Gamma = 71°\ 55'$

Using the formula given on page 1854,

Mean pitch diameter of pinion $d_m = d - F \sin \gamma$
$= 3.2 - 1.5 \times 0.31040$
$= 2.734$ inches
Mean pitch diameter of gear $D_m = D - F \sin \Gamma$
$= 9.8 - 1.5 \times 0.95061$
$= 8.374$ inches

The first step in determining the bearing loads is to compute the tangential, axial, and separating components of the tooth load. Using the formulas on page 1854,

$$W_t = \frac{126{,}050\ P}{nd_m} = \frac{126{,}050 \times 71}{1800 \times 2.734} = 1819 \text{ pounds}$$

$$W_x \text{ for the pinion} = \frac{W_t}{\cos \psi}(\tan \phi \sin \gamma_d + \sin \psi \cos \gamma_d)$$

$$= \frac{1819}{0.81915}(0.36397 \times 0.31040 +$$
$$0.57358 \times 0.95061)$$

$$= 1462 \text{ pounds}$$

$$W_x \text{ for the gear} = \frac{W_t}{\cos \psi}(\tan \phi \sin \gamma_D - \sin \psi \cos \gamma_D)$$

$$= \frac{1819}{0.81915}(0.36397 \times 0.95061 -$$
$$0.57358 \times 0.31040)$$

$$= 373 \text{ pounds}$$

$$W_s \text{ for the pinion} = \frac{W_t}{\cos \psi}(\tan \phi \cos \gamma_d - \sin \psi \sin \gamma_d)$$

$$= \frac{1819}{0.81915}(0.36397 \times 0.95061 -$$
$$0.57358 \times 0.31040)$$

$$= 373 \text{ pounds}$$

$$W_s \text{ for the gear} = \frac{W_t}{\cos \psi}(\tan \phi \cos \gamma_D + \sin \psi \sin \gamma_D)$$

$$= \frac{1819}{0.81915}(0.36397 \times 0.31040 +$$
$$0.57358 \times 0.95061)$$

$$= 1462 \text{ pounds}$$

As explained on Handbook page 1853, the axial thrust load on the bearings is equal to the axial component of the tooth load W_x. Since thrust loads are always taken up at only one mounting point, either bearing A or bearing B must be a bearing capable of taking a thrust of 1462 pounds, and either bearing C or bearing D must be capable of taking a thrust of 373 pounds.

The next step is to determine the magnitudes of the radial loads on the bearings A, B, C, and D. For an *overhung mounted* gear, or pinion, it can be shown, using the principle of moments, that the radial load on bearing A is:

$$R_A = \frac{1}{M} \sqrt{[W_t(L + M)]^2 + [W_s(L + M) - W_x r]^2} \qquad (1)$$

And the radial load on bearing B is:

$$R_B = \frac{1}{M} \sqrt{(W_t L)^2 + (W_s L - W_x r)^2} \qquad (2)$$

For a *straddle mounted* gear or pinion the radial load on bearing C is:

$$R_C = \frac{1}{N + K} \sqrt{(W_t K)^2 + (W_s K - W_x r)^2} \qquad (3)$$

And the radial load on bearing D is:

$$R_D = \frac{1}{N + K} \sqrt{(W_t N)^2 + (W_s N + W_x r)^2} \qquad (4)$$

In these formulas, r is the mean pitch radius of the gear or pinion.

These formulas will now be applied to the gear and pinion bearings in the example. Since an overhung mounting is used for the pinion, Formulas (1) and (2) are used to determine the radial loads on the pinion bearings:

$$R_A = \frac{1}{5}\sqrt{[1819(1.5 + 5)]^2 + [373(1.5 + 5) - 1462 \times 1.367]^2}$$
$$= 2365 \text{ pounds}$$

$$R_B = \frac{1}{5}\sqrt{(1819 \times 1.5)^2 + (373 \times 1.5 - 1462 \times 1.367)^2}$$
$$= 618 \text{ pounds}$$

Since a straddle mounting is used for the gear, Formulas (3) and (4) are used to determine the radial loads on the gear bearings:

$$R_C = \frac{1}{3.5 + 2.5} \sqrt{(1819 \times 2.5)^2 + (1462 \times 2.5 - 373 \times 4.187)^2}$$
$$= 833 \text{ pounds}$$

$$R_D = \frac{1}{3.5 + 2.5} \sqrt{(1819 \times 3.5)^2 + (1462 \times 3.5 + 373 \times 4.187)^2}$$

$$= 1533 \text{ pounds}$$

These radial loads, and the thrust loads previously computed, are then used to select suitable bearings from manufacturers' catalogs.

It should be noted, in applying Formulas (1) to (4), that if both gear and pinion had overhung mountings, then Formulas (1) and (2) would have been used for both; if both gear and pinion had straddle mountings, then Formulas (3) and (4) would have been used for both. In any case, the dimensions and loads for the corresponding member must be used. Also, in applying the formulas, the computed values of W_x and W_s, if they are negative, must be used in accordance with the rules applicable to negative numbers.

PRACTICE EXERCISES FOR SECTION 18
For answers to all practice exercise problems or questions
see Section 22

1. A spur gear of 6 diametral pitch has an outside diameter of 3.3333 inches. How many teeth has it? What is the pitch diameter? What is the tooth thickness measured along the pitch circle?

2. A gear of 6 diametral pitch has 14 teeth. Find the outside diameter, the pitch diameter, and the addendum.

3. When is the 25-degree tooth form standard preferred?

4. What dimension does a gear-tooth vernier caliper measure?

5. What are the principal 20-degree pressure angle tooth dimensions for the following diametral pitches: 4; 6; 8; 18?

6. Give the important 14½-degree pressure angle tooth dimensions for the following circular pitches: ½ inch; ¾ inch; 9⁄16 inch.

7. What two principal factors are taken into consideration in determining the power transmitting capacity of spur gears?

8. The table on Handbook page 1794 shows that a No. 8 formed cutter (involute system) would be used for milling either a 12- or 13-tooth pinion, whereas a No. 7 would be used for tooth numbers from 14 to 16, inclusive. If the pitch is not changed, why is it necessary to use different cutter numbers?

9. Are hobs made in series or numbers for each pitch similar to formed cutters?

10. If the teeth of a gear have a 6/8 pitch, what name is applied to the tooth form?

11. A stub-tooth gear has 8/10 pitch. What do the figures 8 and 10 indicate?

12. What is the module of a gear?

13. Explain the use of the table of chordal thicknesses on Handbook page 1792.

14. Give the dimensions of a 20-degree stub tooth of 12 pitch.

15. What are the recommended diametral pitches for fine-pitch standard gears?

16. What tooth numbers could be used in pairs of gears having the following ratios: 0.2642; 0.9615?

17. What amount of backlash is provided for general purpose gearing and how is the excess depth of cut to obtain it calculated?

18. What diametral pitches correspond to the following modules: 2.75; 4; 8?

19. What is the general rule for determining the face width of bevel gears?

20. Can bevel gears be cut by formed milling cutters?

21. Can the formed cutters used for cutting spur gears also be used for bevel gears?

22. What is the minimum number of teeth recommended for bevel gears of 2 to 1 ratio, assuming that the gears have straight (not spiral) teeth?

23. For spiral bevel gears of 20-degree pressure angle, is the minimum number of teeth recommended for a 2 to 1 ratio smaller than that recommended for straight bevel gears in answer to question 22?

24. What is the pitch angle of a bevel gear?

25. Why is the face cone of a generated bevel gear made parallel to the root cone of its mating gear?

26. When is the term "miter" applied to bevel gears?

27. What is the difference between the terms "whole depth" and "working depth" as applied to gear teeth?

28. Why do pre-shaved gears have a greater dedendum than gears that are finish-hobbed?

29. What AGMA numbers are used to specify gears?

30. Are gear teeth of 8 diametral pitch larger or smaller than teeth of 4 diametral pitch, and how do these two pitches compare in regard to tooth depth and thickness?

31. Where is the pitch diameter of a bevel gear measured?

32. What is the relation between the circular pitch of a wormgear and the linear pitch of the mating worm?

33. From what diameter is the helix angle of a worm calculated?

34. In what respect does the helix angle of a worm differ from the helix angle of a helical or spiral gear?

35. How do the terms "pitch" and "lead," as applied to a worm, compare with the same terms as applied to screw threads?

36. Under what conditions is worm gearing self-locking or incapable of being reversed by driving from wormgear to worm?

37. Why is the outside diameter of a hob for cutting a wormgear somewhat larger than the outside diameter of the worm?

38. How is the number of flutes in a wormgear hob determined?

39. Why are triple, quadruple, or other multiple-threaded worms used when an efficient transmission is required?

40. In designing worm drives having multi-threaded worms, it is common practice to select a number of wormgear teeth that is not an exact multiple of the number of worm threads. Why is this done? When should this practice be avoided?

41. Explain the following terms used in connection with helical or spiral gears: Transverse diametral pitch; normal diametral pitch. What is the relation between these terms?

42. Are helical gear calculations based upon diametral pitch or circular pitch?

43. Can helical gears be cut with the formed cutters used for spur gears?

44. In spiral gearing the tangent of the tooth or helix angle = the circumference ÷ lead. Is this circumference calculated from the outside diameter, the pitch diameter, or the root diameter?

45. What advantages are claimed for gearing of the herring bone type?

CUTTING SPEEDS, FEEDS, AND
MACHINING POWER

HANDBOOK Pages 965–1018

Metal cutting operations such as turning and drilling may not be as productive as they could be unless the material removal rate is at or near the maximum permitted by the available power of the machine. It is not always possible to use the machine's full power owing to limitations imposed by a combination of part configuration, part material, tool material, surface finish and tolerance requirements, coolant employed and tool life. However, even with such restrictions it is practical to find a combination of depth of cut, feed rate, and cutting speed to achieve the best production rate for the job at hand.

The information on Handbook pages 965 to 1018 is useful in determining how to get the most out of machining operations. The tabular data is based upon actual shop experience and extensive testing in machining laboratories. A list of machining data tables is given on page 972, and these tables are referred to in the following.

Most materials can be machined over a wide range of speeds; however, there is usually a narrower spread of speeds within which the most economical results are obtained. This narrower spread is determined by the economical tool life for the job at hand as, for example, when a shorter tool life is tolerable the speed can be increased. On the other hand, if tool life is too short, causing excessive down time, then speed can be reduced to lengthen tool life.

To select the best cutting conditions for machining a part the following procedure may be followed:

1. Select the maximum depth of cut consistent with the job requirements.

2. Select the maximum feed rate that can be used consistent with such job requirements as surface finish and the rigidity of the cutting tool, workpiece, and the machine tool.

3. Select the cutting speed; when possible, use Table 5 to calculate the cutting speed.

This order of selection is based on the laws governing tool life; i.e., the life of a cutting tool is affected most by the cutting speed, then by the feed, and least by the depth of cut.

Using the same order of selection, when very heavy cuts are to be taken, the cutting speed that will utilize the maximum power available on the machine tool can be estimated by using a rearrangement of the machining power formulas given on page 1013. These formulas are used together with those given on pages 973 and 996, which are used when taking ordinary cuts, as well as heavy cuts. Often the available power on the machine will limit the size of the cut that can be taken, in which case the maximum depth of cut and feed should be used and the cutting speed adjusted to utilize the maximum available power. When the cutting speed determined in this manner is equal to or less than recommended, the maximum production and the best possible tool life will be achieved. When the estimated cutting speed is greater than recommended, the depth of cut or feed may be increased, but the cutting speed should not be increased beyond that which will provide a reasonable tool life.

Example 1:—An ASTM Class 25 (160–180 HB) gray iron casting is to be turned on a geared head lathe using a cemented carbide cutting tool. The heaviest cut will be .250 inch (6.35 mm) deep, taken on an 8-inch (203.2 mm) diameter of the casting; a feed rate of .020 in./rev (0.51 mm/rev) is selected for this cut. Using both inch and metric units calculate the spindle speed of the lathe and estimate the power required to take this cut. (Terms are defined on Handbook pages 973 and 1013–1017, and in the tables.)

Inch units:

$$V_o = 350 \text{ fpm (Table 4)} \qquad C = .90 \text{ (Table 26)}$$
$$F_f = .80 \quad \text{(Table 5)} \qquad Q = 12Vfd \text{ (Table 29)}$$
$$F_d = .91 \quad \text{(Table 5)} \qquad W = 1.30 \text{ (Table 27)}$$
$$K_p = .52 \quad \text{(Table 24)} \qquad E = .80 \text{ (Table 28)}$$

$$V = V_o F_f F_d = 350 \times .80 \times .91 = 255 \text{ fpm}$$

$$N = \frac{12V}{\pi D} = \frac{12 \times 255}{\pi \times 8} = 122 \text{ rpm}$$

$$Q = 12Vfd = 12 \times 255 \times .020 \times .250 = 15.3 \text{ in.}^3/\text{min}$$

$$P_m = \frac{K_p CQW}{E} = \frac{.52 \times .90 \times 15.3 \times 1.30}{.80} = 11.64 \text{ hp}$$

Metric units:

$V_o = 350 \times 0.3 = 105$ m/min (Table 4)

$F_f = 0.80$ (Table 5) $C = 0.90$ (Table 26)

$F_d = 0.91$ (Table 5)

$\qquad\qquad\qquad\qquad Q = \dfrac{V}{60} fd$ (Table 29)

$K_p = 1.42$ (Table 24) $W = 1.30$ (Table 27)

$\qquad\qquad\qquad\qquad E = 0.80$ (Table 28)

(The inch and corresponding metric values of the feed and depth of cut factors, F_f and F_d, are always equal.)

$$V = V_o F_f F_d = 105 \times 0.80 \times 0.91 = 76.4 \text{ fpm}$$

$$N = \frac{1000V}{\pi D} = \frac{1000 \times 76.4}{\pi \times 203.2} = 120 \text{ rpm}$$

$$Q = \frac{V}{60} fd = \frac{76.4}{60} \times 0.51 \times 6.35 = 4.12 \text{ cm}^3/\text{s}$$

$$P_m = \frac{K_p C Q W}{E} \frac{1.42 \times 0.90 \times 4.12 \times 1.30}{0.80} = 8.56 \text{ kW}$$

Example 2:—If the lathe in Example 1 has only a 10 hp motor, estimate the cutting speed and spindle speed that will utilize the maximum available power. Use inch units only.

$$Q_{max} = \frac{P_m E}{K_p C W} = \frac{10 \times .80}{.52 \times .90 \times 1.30} \quad \left(P_m = \frac{K_p C Q W}{E} \right)$$

$$= 13.15 \text{ in.}^3/\text{min}$$

$$V = \frac{Q_{max}}{12 fd} = \frac{13.5}{12 \times .020 \times .250} \quad (Q = 12Vfd)$$

$$= 219 \text{ fpm}$$

$$N = \frac{12V}{\pi D} = \frac{12 \times 219}{\pi \times 8} = 105 \text{ rpm}$$

Example 3:—A slab milling operation is to be perfomed on 120–140 HB AISI 1020 steel using a 3-inch diameter high speed steel plain milling cutter having 8 teeth. The width of this cut is 2 inches;

the depth is .250 inch, and the feed rate is .004 in./tooth. Estimate the power at the motor required to take this cut.

$V = 110$ fpm (Table 11) $Q = f_m wd$ (Table 29)
$K_P = .69$ (Table 23) $W = 1.10$ (Table 27)
$C = 1.25$ (Table 26) $E = .80$ (Table 28)

$$N = \frac{12V}{\pi D} = \frac{12 \times 110}{\pi \times 3} = 140 \text{ rpm}$$

$$f_m = f_t \, n_t \, N = .004 \times 8 \times 140 = 4.5 \text{ in./min}$$

$$Q = f_m \, wd = 4.5 \times 2 \times .250 = 2.25 \text{ in.}^3/\text{min}$$

$$P_m = \frac{K_p CQW}{E} = \frac{.69 \times 1.25 \times 2.25 \times 1.10}{.80} = 2.67 \text{ hp}$$

Example 4:—A 16-inch diameter cemented carbide face milling cutter having 18 teeth is to be used to take a 14-inch wide and .125-inch deep cut on an H12 tool steel die block having a hardness of 250–275 HB. The feed used will be .008 in./tooth, and the milling machine has a 20 hp motor. Estimate the cutting speed and the spindle speed to be used which will utilize the maximum horsepower available on the machine.

$K_p = .98$ (Table 25) $W = 1.25$ (Table 27)
$C = 1.08$ (Table 26) $E = .80$ (Table 28)
$Q = f_m wd$ (Table 29)

$$Q_{max} = \frac{P_m E}{K_p CW} = \frac{20 \times .80}{.98 \times 1.08 \times 1.25} \left(P_m = \frac{K_p CQW}{E} \right)$$

$$= 12 \text{ in.}^3/\text{min}$$

$$f_m = \frac{Q_{max}}{wd} = \frac{12}{14 \times .125} \quad (Q = f_m wd)$$

$$= 6.86 \text{ in./min.; use } 7 \text{ in./min}$$

$$N = \frac{f_m}{f_t n_t} = \frac{7}{.008 \times 18} \quad (f_m = f_t \, n_t \, N)$$

$$= 48.6 \text{ rpm; use } 50 \text{ rpm}$$

$$V = \frac{\pi D N}{12} = \frac{\pi \times 16 \times 50}{12} = 209 \text{ fpm}$$

Formulas for estimating the thrust, torque, and power for drilling are given on page 1016. Thrust is the force required to push or feed the drill when drilling. This force can be very large. It is sometimes helpful to have knowledge of the magnitude of this force and the torque exerted by the drill when designing drill jigs or work-holding fixtures; it is essential to have this information as well as the power required to drill when designing machine tools on which drilling operations are to be performed. In the ordinary shop it is often helpful to be able to estimate the power required to drill larger holes in order to determine if the operation is within the capacity of the machine to be used.

Example 5:—Estimate the thrust, torque, and power at the motor required to drill a ¾-inch diameter hole in a part made from AISI 1117 steel, using a conventional twist drill and a feed rate of .008 in./rev.

$$
\begin{aligned}
K_d &= 12{,}000 \text{ (Table 30)} & B &= 1.355 & \text{(Table 33)} \\
F_f &= .021 \text{ (Table 31)} & J &= .030 & \text{(Table 33)} \\
F_T &= .794 \text{ (Table 32)} & E &= .80 & \text{(Table 28)} \\
F_M &= .596 \text{ (Table 32)} & W &= 1.30 & \text{(Table 27)} \\
A &= 1.085 \text{ (Table 33)} & V &= 110 \text{ fpm} & \text{(Table 17)}
\end{aligned}
$$

$$
\begin{aligned}
T &= 2\,K_d\,F_f\,F_T\,BW + K_d d^2 JW \\
&= 2 \times 12{,}000 \times .021 \times .794 \times 1.355 \times 1.30 \\
&\quad + 12{,}000 \times .750^2 \times .030 \times 1.30 \\
&= 968 \text{ lb}
\end{aligned}
$$

$$
\begin{aligned}
M &= K_d\,F_f\,F_M\,AW \\
&= 12{,}000 \times .021 \times .596 \times 1.085 \times 1.30 \\
&= 212 \text{ in. lb}
\end{aligned}
$$

$$
\begin{aligned}
N &= \frac{12V}{\pi D} = \frac{12 \times 110}{\pi \times .750} \\
&= 560 \text{ rpm}
\end{aligned}
$$

$$
\begin{aligned}
P_c &= \frac{MN}{63{,}025} = \frac{212 \times 560}{63{,}025} \\
&= 1.88 \text{ hp}
\end{aligned}
$$

$$
\begin{aligned}
P_m &= \frac{P_c}{E} = \frac{1.88}{.80} \\
&= 2.35 \text{ hp}
\end{aligned}
$$

PRACTICE EXERCISES FOR SECTION 19

For answers to all practice exercise problems or questions
see Section 22

1. Calculate the spindle speeds for turning ½-inch and 4-inch bars
made from the following steels, using a high speed steel cutting tool
and the cutting conditions given as follows:

Steel Designation	Feed, in./rev	Depth of Cut, in.
AISI 1108, Cold Drawn	.012	.062
12L13, 150–200 HB	.008	.250
1040, Hot Rolled	.015	.100
1040, 375–425 HB	.015	.100
41L40, 200–250 HB	.015	.100
4140, Hot Rolled	.015	.100
O2, Tool Steel	.012	.125
M2, Tool Steel	.010	.200

2. Calculate the spindle speeds for turning 6-inch diameter sections of the following materials, using a cemented carbide cutting
tool and the cutting conditions given below:

Material	Feed, in./rev	Depth of Cut, in.
AISI 1330, 200 HB	.030	.150
201 Stainless Steel, Cold Drawn	.012	.100
ASTM Class 50 Gray Cast Iron	.016	.125
6A1-4V Titanium Alloy	.018	.188
Waspaloy	.010	.062

3. A 200 HB AISI 1030 forged steel shaft is being turned at a
constant spindle speed of 400 rpm, using a cemented carbide cutting
tool. The as-forged diameters of the shaft are 1½, 3, and 4 inches.
Calculate the cutting speed (fpm) at these diameters and check to
see if they are within the recommended cutting speed.

4. A 75 mm diameter bar of cold drawn wrought aluminum is to
be turned with a high speed steel cutting tool, using a cutting speed
of 180 m/min. Calculate the spindle speed that should be used.

5. Calculate the spindle speed required to mill a 745 nickel silver
part using a ½-inch end milling cutter.

6. An AISI 4118 part having a hardness of 200 HB is to be machined on a milling machine. Calculate the spindle speeds for each of the operations below and the milling machine table feed rates for Operations 1 and 2.

1. Face mill top surface, using an 8-inch diameter cemented carbide face milling cutter having 10 teeth. (Use f_t = .008 in./rev.)

2. Mill ¼ in. deep slot, using a ¾-inch diameter two fluted high speed steel end milling cutter.

3. Drill a ²³⁄₆₄ inch hole.

4. Ream the hole ⅜ inch, using HSS reamer.

7. A 3-inch diameter high speed steel shell end milling cutter having 12 teeth is used to mill a piece of D2 high carbon, high chromium cold work tool steel having a hardness of 220 HB. The spindle speed used is 75 rpm and the milling machine table feed rate is 10 in./min. Check the cutting conditions with respect to the recommended values, and make recommendations for improvements, if possible.

8. A 100–150 HB low carbon steel casting is to be machined with a 12-inch diameter cemented carbide face milling cutter having 14 teeth, using a spindle speed of 60 rpm and a table feed rate of 5 in./min. Check these cutting conditions and recommend improvements, if possible.

9. Estimate the power at the cutter and at the motor required to turn 210 HB AISI 1040 steel in a geared head lathe, using a depth of cut of .125 in., a feed of .015 in./rev., and a cutting speed of 300 fpm. (Let E = .80.)

10. A 165 HB A286 high temperature alloy, or superalloy, is to be turned on a 3 hp geared head lathe using a cemented carbide cutting tool. The depth of cut selected is .100 inch and the feed is .020 in./rev. Estimate the cutting speed that will utilize the maximum power available on the lathe.

11. An AISI 8642 steel having a hardness of 210 HB is to be milled with a 6-inch diameter cemented carbide face milling cutter having 8 teeth on a 10 hp milling machine. The depth of cut is to be .200 inch, the width is 4 inches and the feed is to be .010 in./tooth. Estimate the cutting speed that will utilize the maximum power available on the machine.

12. Estimate the thrust, torque and power at the motor required to drill 200 HB steel using the following drill sizes, feeds, and spindle speeds.

Drill Size	Feed	Spindle Speed
¼ in.	0.0005 in./rev	1500 rpm
½ in.	0.002 in./rev	750 rpm
1 in.	0.008 in./rev	375 rpm
19 mm	0.15 mm/rev	500 rpm

13. Estimate the thrust, torque, and power at the motor for the 1-inch drill in Question 12 if the drill is ground to have a split point.

14. Describe the general characteristics of high speed steels that make them suitable for use as cutting tool materials.

15. What guidelines should be followed in selecting a grade of cemented carbide?

16. How does the cutting speed, feed, and depth of cut influence tool life?

17. List the steps for selecting the cutting conditions in their correct order and explain why.

18. What are the advantages of coated carbides and how should they be used?

19. Name the factors that must be considered when selecting a cutting speed for tapping.

20. Name the factors that must be considered when selecting a cutting speed for cutting threads with a single point tool.

21. Why is it important to calculate the table feed rate for milling?

22. Name the factors that affect the basic feed rate for milling.

23. When should the power required to take a cut be estimated? Why?

24. Name the factors that affect the power constant, K_p. This constant is unaffected by what?

25. Why is it necessary to have a separate method for estimating the drilling thrust, torque, and power?

SECTION 20

NUMERICAL CONTROL

Handbook Pages 1117–1168

Numerical control (NC) is defined by the Electronic Industries Association as "a system in which actions are controlled by the direct insertion of numerical data at some point. The system must automatically interpret at least some portion of these data." Applied to machine tools, NC is used to tell the unit what to do in such explicit detail that it can produce a component part or parts in a completely automatic cycle without intervention from the operator. This cycle may extend from loading of a raw casting or other workpiece through unloading of a finished component ready for assembly, and can be repeated precisely, as often as required. An important aspect of NC is that machines so equipped can often be set up to process even single components economically.

Apart from systems that are designed to load, locate, and clamp the part to be machined, and to select the tool and the spindle speed to be used, for instance, NC installations use programs designed to control movements of the cutting edge of the tool relative to the work (or the work relative to the tool). These machining control instructions, called part programs, may be put together by a machine operator with a push-button panel on the machine if the part is simple, or they may be written in an engineering office, often with the aid of a computer. Some part programs may provide for simply moving the tool or workpiece from one position, at which a fixed machining cycle (known as a subroutine or subprogram) is to be performed, to other positions where the same cycle is to be repeated and triggering the subroutine at each position. Such a program is called point-to-point positioning. There are subroutines for drilling, reaming, counterboring, and tapping, for which tools will be inserted into, clamped, and removed from the spindle automatically.

Other, more complex, programs may be written to cause the workpiece to move past the cutting tool in a series of curves, to generate contoured surfaces on the work. Such a program is called a continuous-path or contouring program. In the associated machining operation, the movements of the table carrying the workpiece along (usually) two axes, and (sometimes) of the spindle head holding the cutter along one axis, are coordinated by electronic signals in a binary digital code that are converted to DC or AC power and fed continuously to controllers connected to the units powering the slides. Measuring equipment attached to each leadscrew or slide provides continuous feedback information of the slide position to the control system for comparison with the command program.

Information in the Handbook, pages 1117 to 1168, is arranged in alphabetic sequence for ease of reference and, because of the complexity of the subject, depends mainly on definitions to explain the various aspects. Most attention is paid to the use of the Automatic Programmed Tool (APT) language in part programming, and examples of typical computational and geometric programs are discussed. For instance, the APT language can be used to specify the four arithmetical operations and the exponential and trigonometric computations used in many algebraic formulas. The APT language visualizes the part program as if it were designed to move the tool past a stationary workpiece, but the formulas for generation of the required shapes most often are translated by the control system into movements of the slides to carry the workpiece past the cutting tool.

Point-to-point Programming.—As an example of the use of NC for point-to-point part programs, consider the rectangular plate shown in Fig. 1, in which it is required to machine eight holes as shown. Dimensions for the positions of the holes are here provided in terms of their distances from X and Y axes, which are conveniently located at a central point on the part. This positioning information is easily transferred to the punched paper tape or other means used to feed it to the machine. Instructions for the tooling to be loaded into the spindle for the work to be performed are also included in the part program, in accordance with the special codes, many of which are listed in the Handbook. The hole location information in the table, Fig. 1, is entered in a part programming manuscript, together with coded details such as spindle speed and feed rates, and is subsequently punched into a tape that will be read by the NC machine when the machining work is started.

Continuous-path Programming.—Surfaces at angles to the axes, and curved surfaces, are produced by continuous-path, or contouring, programs. These programs coordinate two or more machine motions

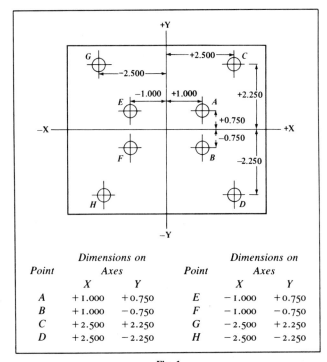

Point	Dimensions on Axes		Point	Dimensions on Axes	
	X	Y		X	Y
A	+ 1.000	+ 0.750	E	− 1.000	+ 0.750
B	+ 1.000	− 0.750	F	− 1.000	− 0.750
C	+ 2.500	+ 2.250	G	− 2.500	+ 2.250
D	+ 2.500	− 2.250	H	− 2.500	− 2.250

Fig. 1.

simultaneously and precisely, so that the movement of the workpiece relative to the cutting tool generates the required curved shape. Angular shapes are generated by straight-line or linear interpolation programs that coordinate movements of two slides to produce the required angle. Circular arcs can be generated by means of a circular interpolation program that controls the slide movements automatically to produce the curved outline. Arcs that are not circular generally must be broken down into a sequence of straight-line segments. Surfaces generated by this method can be held within tolerance by using a large number of segments closely spaced together.

For example, in programming the movement of a cutter, relative to the workpiece, along the curved line shown in the diagram, Fig. 2, it is first necessary to indicate that the cutter is to move in a clockwise and circular path by inserting into the tape the code GO2. Next, the movements along the X and Y axes, which define the component lengths of the arc, are inserted. In Fig. 2, the X movement is $+1.4000$ in. and the Y movement is -1.5000 in. The I dimension of 0.7500 in. parallel to the X axis is the horizontal distance of point A from the arc center and is next included in the program. The vertical distance J of 1.9000 in. from the arc center to the circle is next entered, and the feed rate also must be punched in.

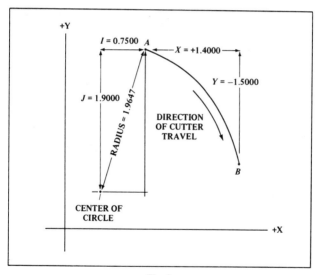

Fig. 2.

PRACTICE EXERCISES FOR SECTION 20

For answers to all practice exercise problems or questions see
Section 22

1. List five or more machine tools on which point-to-point programming is used.

2. List five or more applications of continuous-path or contouring programs.

3. Give some reasons why NC machines are being used increasingly.

4. Which of the following applications of NC is the most used?
(a) Grinding (b) Turning (c) Broaching

5. A _____ is a rotary device used to feed signals to the control system to close the servo loop of an NC installation.

6. CNC systems are far superior to their hardwire predecessors. Name several advantages of CNC systems.

7. What purpose is served by the two feedbacks in an NC servo system?

8. If a stepping motor connected directly to a leadscrew rotates 1.8 degrees per pulse, how far would a 5-pitch leadscrew move a slide if the motor received 254 pulses?

9. The RS-244-B Standard states that words are made up of characters. (True, False.)

10. With a CNC system, the *F* or feedrate word is most commonly described as (a) A ratio of ipm feed divided by the distance moved. (b) Directly in ipm. (c) By the formula of the "magic three."

11. The word that identifies a block is called a _____ _____ .

12. The word address letter for the velocity of a slide on an NC machine is _____ .

13. What is the difference between cutter offset and cutter compensation?

14. Considering the plane formed by the intersection of the *X* and *Y* axes, how many quadrants may be available to a part programmer?

15. Circular interpolation reduces the number of straight-line

segments required to be calculated when a machine is moving about a circular arc. (True, False.)

16. With most control systems, how many blocks would be needed to move around a complete circle (360 degrees) when circular interpolation is used?

17. In the first column below are shown the various subroutines or canned cycles. In the second column are some preparatory codes. Match the functions with the codes.

a.	Drill plus dwell	1.	G89
b.	Deep hole drill	2.	G81
c.	Boring, spindle rotating on withdrawal at feedrate	3.	G85
d.	Drill	4.	G84
e.	Tapping	5.	G82
f.	Boring, spindle rotating on withdrawal at feedrate plus dwell	6.	G83

18. A parametric subroutine is used exclusively for describing the path around the outside of a part. (True, False.)

19. Computer-aided part programming refers to the assistance offered by the computer within the CNC system. (True, False.)

20. The most common media for transmitting data and instructions to an NC system is (a) Floppy disc. (b) Magnetic tape. (c) Punched tape.

21. Name the three surfaces involved in an APT move.

22. Two of the three surfaces in APT appear as lines when viewed from directly above. What are these surfaces?

23. What is a G word?

24. What is an APT start-up statement?

25. Explain the rule that describes the orientation and directions of the motions of slides and spindles on a machine tool.

SECTION 21

GENERAL REVIEW QUESTIONS

For answers, see Section 22

1. If a regular polygon of 20 sides is to have an area of 100 square inches what formula may be used to calculate the length of one side of the polygon?

2. What does the number of a Jarno taper indicate?

3. What is the general rule for determining the direction in which to apply tolerances?

4. Why is 1 horsepower equivalent to 33,000 foot-pounds of work per minute? Why not 30,000 or some other number?

5. What is the chief element in the composition of babbitt metals?

6. If the pitch of a stub tooth gear is 8/10, what is the tooth depth?

7. What does the figure 8 mean if the pitch of a stub tooth gear is 8/10?

8. Explain how to determine the diametral pitch of a spur gear from a sample gear.

9. If a sample gear is cut to circular pitch, how can this pitch be determined?

10. What gage is used for seamless tubing and does it apply to all metals?

11. How does the strength of iron wire rope compare with steel rope?

12. Is the friction between two bearing surfaces proportional to the pressure?

13. If the surfaces are well lubricated, upon what does frictional resistance depend?

14. What is the general rule for subtracting a negative number from a positive number? For example, $8 - (-4) = ?$

15. Is one meter longer than one yard?

16. On Handbook page 2484 two of the equivalents of one horse-power-hour are: 1,980,000 foot-pounds, and 2.64 pounds of water evaporated at 212 degrees F. How is this relationship between work and heat established?

17. Is "extra strong" and "double extra strong" wrought or steel pipe larger in diameter than standard weight pipe?

18. In the design of plain bearings what is the general relationship between surface finish and hardness of journal?

19. Are the nominal sizes of wrought or steel pipe ever designated by giving the outside diameter?

20. What are the advantages of plastic pipe?

21. Will charcoal ignite at a lower temperature than dry pine?

22. What general classes of steel are referred to as "stainless"?

23. What kind of steel is known as Bessemer screw stock?

24. Does the nominal length of a file include the tang? For example, is a 12-inch file 12 inches long over all?

25. Is steel heavier than cast iron?

26. Is there any alloy that will melt at a temperature below the boiling point of water?

27. What is the specific gravity (a) of solid bodies, (b) of liquids, (c) of gases?

28. A system of four-digit designations for wrought aluminum and aluminum alloys was adopted by The Aluminum Association in 1954. What do the various digits signify?

29. What alloys are known as "red brass" and how do they compare with "yellow brass"?

30. What is the difference between adiabatic expansion or compression and isothermal expansion or compression?

31. Are the sizes of all small twist drills designated by numbers?

32. Why are steel tools frequently heated in molten baths for hardening them?

33. In hardening tool steel, what is the best temperature for refining the grain of the steel?

34. In cutting a screw thread on a tap assume that the pitch is to be increased from 0.125 inch to 0.1255 inch to compensate for shrinkage in hardening. How can this be done?

35. What is the general rule for reading a vernier scale (*a*) for linear measurements; (*b*) for angular measurements?

36. The end of a shaft is to be turned to a taper of ¾ inch per foot for length of 5 inches without leaving a shoulder at the end of the cut. How is the diameter of the small end determined?

37. Is there a simple way of converting the function of 90° plus an angle to the function of the angle itself?

38. What decimal part of a degree is 53 minutes?

39. If $10x - 5 = 3x + 16$, what is the value of x?

40. Approximately what angle is required for a cone clutch to prevent either slipping or excessive wedging action?

41. What is the coefficient of friction?

42. Is Stub's steel wire gage used for the same purpose as Stub's iron wire gage?

43. Why are some ratchet mechanisms equipped with two pawls of different lengths?

44. How should a leather belt be applied to pulleys?

45. If a quarter-turn belt drive cannot be avoided, how should the driving and driven pulleys be aligned?

46. Is the ultimate strength of a crane or hoisting chain equal to twice the ultimate strength of the bar or rod used for making the links?

47. How would you determine the size of chain required for lifting a given weight?

48. If a shaft 3½ inches in diameter is to be turned at a cutting speed of 90 feet per minute, what number of revolutions per minute will be required?

49. In lapping by the "wet method," what kind of lubricant is preferable (*a*) with a steel lap, (*b*) with a cast-iron lap?

50. What is the meaning of the terms right-hand and left-hand as applied to helical or spiral gears, and how is the "hand" of the gear determined?

51. Are mating helical or spiral gears always made to the same hand?

52. How would you determine the total weight of 100 feet of 1½-inch standard weight pipe?

53. What is the difference between casehardening and pack-hardening?

54. What is the nitriding process of heat-treating steel?

55. What is the difference between single-cut and double-cut files?

56. For general purposes what is the usual height of work benches?

57. What do the terms "major diameter" and "minor diameter" mean as applied to screw threads in connection with the American Standard?

58. Is the present S. A. E. Standard for screw threads the same as the Unified and American Standard?

59. Does the machinability of steel depend only upon its hardness?

60. Is there any direct relationship between the hardness of steel and its strength?

61. What is the millimeter equivalent of $^{33}/_{64}$ths of an inch?

62. How is the sevolute function of an angle calculated?

63. What is the recommended cutting speed in feet per minute for turning normalized AISI 4320 alloy steel with a Bhn hardness of 250, when using a carbide tool?

64. The diametral pitch of a spur gear equals the number of teeth divided by pitch diameter. Is the diametral pitch of the cutter or hob for a helical or spiral gear determined in the same way?

65. What are the design recommendations for carburized gears and pinions where considerable impact loads are to be resisted?

66. Are the symbols for dimensions and angles used in spline calculations the same for both inch-dimension and metric module involute splines?

67. What kind of bearing surface and tool insert rake are provided by an indexable insert tool holder?

68. Is it necessary in making ordinary working drawings of gears to lay out the tooth curves?

69. In milling plate cams on a milling machine, how is the cam rise varied other than by changing the gears between the dividing head and feed screw?

70. How is the angle of the dividing head spindle determined for milling plate cams?

71. How is the center-to-center distance between two gears determined if the number of teeth and diametral pitch are known?

72. How is the center-to-center distance determined in the case of internal gears?

73. In the failure of riveted joints, rivets may fail through one or two cross-sections or by crushing. How may plates fail?

74. What gage is used in England to designate wire sizes?

75. What is a Prony brake?

76. What is the advantage of a Prony brake or other form of dynamometer for measuring power?

77. If a beam supported at each end is uniformly loaded throughout its length, will its load capacity exceed that of a similar beam loaded at the center only?

78. Is there any relationship between Brinell hardness and tensile strength of steel?

79. Is the outside diameter of a 2-inch pipe about 2 inches?

80. The hub of a lever 10 inches long is secured to a 1-inch shaft by a taper pin. If the maximum pull at the end of the lever equals 60 pounds, what pin diameter is required? (Give mean diameter or diameter at center.)

81. What are the two laws that form the basis of all formulas relating to the solution of triangles?

82. What are the sine and the cosine of the angle 45 deg.

83. How is the pressure of water in pounds per square inch determined for any depth?

84. When calculating the basic load rating for a unit consisting of 2 bearings mounted in tandem, is the rated load of the combination equal to 2 times the capacity of a single bearing?

85. If a machine producing 50 parts per day is replaced by a machine that produces 100 parts per day, what is the percentage of increase?

86. If production is decreased from 100 to 50, what is the percentage of reduction?

87. What kind of steel is used ordinarily for springs in the automotive industry?

88. What is the heat-treating process known as "normalizing"?

89. What important standards apply to electric motors?

90. Is there an American standard for section linings to represent different materials on drawings?

91. Is the taper per foot of the Morse standard uniform for all numbers or sizes?

92. Is there more than one way to remove a tap that has broken in the hole during tapping?

93. The center-to-center distance between two bearings for gears is to be 10 inches, with a tolerance of 0.005 inch. Should this tolerance be (a) unilateral and plus, (b) unilateral and minus, (c) bilateral?

94. How are the available pitch diameter tolerances for Acme screw threads obtained?

95. On Handbook page 1172, there is a rule for determining the pressure required for punching circular holes into steel sheets or plates. Why is the product of the hole diameter and stock thickness multiplied by 80 to obtain the approximate pressure in tons?

96. What gage is used in the United States for cold-rolled sheet steel?

97. What gage is used for brass wire, and is the same gage used for brass sheets?

98. Is the term "babbitt metal" applied to a single composition?

99. Next to tin, what are the chief elements in high-grade babbitt metal?

100. How many bars of stock 20 feet long will be needed to make 20,000 dowel-pins 2 inches long, if the tool for cutting them off is 0.100 inch wide?

101. What is the melting point and weight per cubic inch of cast iron; steel; lead; copper; nickel; tin?

102. What lubricant is recommended for machining aluminum?

103. What relief angles are recommended for cutting copper, brass, bronze and aluminum?

104. Why is stock annealed between drawing operations in producing parts in drawing dies?

105. When is it advisable to mill screw threads?

106. How does a fluted chucking reamer differ from a rose chucking reamer?

107. What kind of material is commonly used for gage blocks?

108. What grade of gage blocks are used as shop standards?

109. What is the "lead" of a milling machine?

110. The table on Handbook page 1723 shows that a lead of 9.625 inches will be obtained if the numbers of teeth in the *driven* gears are 44 and 28, and the numbers of teeth on the *driving* gears 32 and 40. Prove that this lead of 9.625 inches is correct.

111. Use the prime number and factor table beginning on Handbook page 3 to reduce the following fractions to their lowest terms: $210/462$; $2765/6405$; $741/1131$.

112. If a bevel gear and a spur gear each have 30 teeth of 4 diametral pitch, how do the tooth sizes compare?

113. For what types of work are the following machinists' files used: (*a*) flat files? (*b*) half round files? (*c*) hand files? (*d*) knife files? (*e*) general purpose files? (*f*) pillar files?

114. Referring to the illustration on Handbook page 686, what is the dimension x over the rods used for measuring the dovetail slide if a is 4 inches, angle α is 60 degrees, and the diameter of the rods used is ⅝ inch?

115. Determine the diameter of the bar or rod for making the links of a single chain required to lift safely a load of 6 tons.

116. Why will a helical gear have a greater tendency to slip on an arbor while the teeth are being milled than when milling a straight tooth gear?

117. How does a collapsing tap reduce the time for tapping?

118. When is a removable or "slip" bushing used in a jig?

119. What are the relative ratings and properties of an H43 molybdenum high speed tool steel?

120. What systematic procedure may be used in designing a roller chain drive to meet certain requirements as to horsepower, center distance, etc.?

121. In the solution of oblique triangles having 2 sides and the angle opposite one of the side known, it is possible to have no solution or more than one solution. Under what condition will there be no solution?

122. What gear steels would you use (1) for casehardened gears? (2) for fully hardened gears? (3) for gears which are to be machined after heat-treatment?

123. Is it practicable to tap holes and obtain (1) Class 2B fits? (2) Class 3B fits?

124. What is the maximum safe operating speed of an organic bonded Type 1 grinding wheel when used in a bench grinder?

125. What is the recommended type of diamond wheel and abrasive specification for internal grinding?

126. Is there a standard direction of rotation for all types of nonreversing electric motors?

127. Anti-friction bearings are normally grease lubricated. Is oil ever used? If so, when?

128. In the example on page 1671, the side relief angle at the leading edge of the single point acme thread cutting tool was calculated to be 19.27°, or 19°16′, which provides an effective relief angle (a_e) between the flank of the tool and the side of the thread of 10° at the minor diameter. What is the effective relief angle of this tool at the pitch diameter (E) and at the major diameter (D)? The pitch diameter of the thread is .900 inch, the major diameter is 1.000 inch, and the lead of the thread is .400 inch.

129. Helical flute milling cutters having eccentric relief are known to provide better support of the cutting edge than cutters ground with straight or concave relief. For a 1-inch diameter milling cutter

having a 35-degree helix angle, what is the measured indicator drop according to the methods described beginning on page 791 if the radial relief angle is to be 7°?

130. On page 2127 in the Handbook, Table 13 shows that TFE fabric bearings have a load capacity of 60,000 pounds per square inch. Also shown in the table is a PV limit of 25,000 for this material. What is the maximum surface speed in feet per minute that this material can operate at when the load is 60,000 psi?

131. Is there a standard for shaft diameter and housing bore tolerance limits applying to rolling element bearings?

132. In designing an aluminum bronze plain bearing what hardness should the steel journal have?

133. Steel balls are usually sold by the pound. How many pounds will provide 100 balls of $^{13}/_{32}$-inch diameter carbon steel?

134. When using silent chains why should a hunting tooth ratio be employed whenever possible?

135. If a 3AM1-18 steel retaining ring were used on a rotating shaft, what is the maximum allowable speed of rotation?

136. What procedure applies to 3-wire measurements of Acme threads when the lead angle is greater than 5 degrees?

137. Twelve 1½ diameter rods are to be packed in a tube. What is the minimum inside diameter of the tube?

SECTION 22

ANSWERS TO PROBLEMS AND QUESTIONS

All references are to HANDBOOK page numbers

Section Number	Number of Question	Answers (Or where information is given in Handbook)
1	1	78.54 mm²; 31.416 mm
	2	4.995 or 5, approx.
	3	3141.6 mm²
	4	127.3 psi
	5	1.27
	6	1.5708
	7	8 hours, 50 minutes
	8	2450.448 pounds
	9	2¹⁄₁₆ inches
	10	7 degrees, 10 minutes
	11	Yes. The *x*, *y*, coordinates given in the tables of Jig Boring coordinates, pages 953 to 961, may be used
2	1	Page 964
	2	(a) 0.043 inch (b) 0.055 inch (c) 0.102 inch
	3	0.336 inch
	4	2.796 inches
	5	4.743 inches
	6	4.221 feet
	7	Page 70
	8	740 gallons, approximately
	9	Formula on page 63
	10	Formula on page 63
	11	Formulas on page 63

Section Number	Number of Question	Answers (Or where information is given in Handbook)
3	1	(a) 104 horsepower: (b) If reciprocal is used, $H = 0.33\ D^2SN$.
	2	65 inches
	3	5.74 inches
	4	Side $s = 5.77$ inches; diagonal $d = 8.165$ inches, and volume = 192.1 cubic inches; $d = 8.165$ inches
	5	91.0408 square inches
	6	4.1888 and 0.5236
	7	59.217 cubic inches
	8	Page 1
	9	$a = \dfrac{2A}{h} - b$
	10	$r = \sqrt{R^2 - \dfrac{s^2}{4}}$
	11	$a = \sqrt{\dfrac{\left(\dfrac{P}{\pi}\right)^2}{2} - b^2}$
	12	$\sin A = \sqrt{1 - \cos^2 A}$
	13	$a = \dfrac{b \times \sin A}{\sin B}$; $b = \dfrac{a \times \sin B}{\sin A}$; $\sin A = \dfrac{a \times \sin B}{b}$; $\sin B = \dfrac{b \times \sin A}{a}$
4	1	It is easier and more accurate to use a handheld calculator for multiplication and division.
	2	Yes. The use of Continued or Conjugate Fractions as described in Handbook pages 12–14
	3	Yes. See the formulas for Friction Brakes beginning on Handbook page 2228.

Section Number	Number of Question	Answers (Or where information is given in Handbook)
5	1	8001.3 cubic inches
	2	83.905 square inches
	3	69.395 cubic inches
	4	1.299 inches
	5	22.516 cubic inches
	6	8 inches
	7	0.0276 cubic inches
	8	4.2358 inches
	9	1.9635 cubic inches
	10	410.5024 cubic inches
	11	26.4501 square inches
	12	Radius 1.4142 inches; area, 0.43 square inch
	13	Area, 19.869 square feet; volume, 10.2102 cubic feet
	14	Area, 240 square feet; volume, 277.12 cubic feet
	15	11.3137 inches
	16	41.03 gallons
	17	17.872 square inches
	18	1.032 inches
	19	40 cubic inches
	20	Table page 73
	21	Table page 73
	22	5.0801 inches
	23	4 inches; 5226 inches
6	1	Page 35
	2	Page 35
	3	Page 35
	4	Page 35
	5	Page 36
	6	Page 36
	7	Page 36
	8	Page 36
	9	Page 37
	10	Page 37

Section Number	Number of Question	Answers (Or where information is given in Handbook)
	11	Page 37
	12	Page 37
	13	Page 39
	14	Page 40
	15	Page 38
	16	Page 38
	17	Page 38
	18	Page 38
6	19	Page 38
	20	Page 39
	21	Page 39
	22	Page 39
	23	Page 40
	24	Page 40
	25	Page 40
	26	Page 40
	1	See page 81
	2	In any right-angle triangle having an acute angle of 30 degrees, the side opposite that angle equals 0.5 × hypotenuse.
	3	Sine = 0.31634; tangent = 0.51549; cosine = 0.83942
	4	Angles equivalent to tangents are 27° 29′ 24″ and 7° 25′ 16″; angles equivalent to cosines are 86° 5′ 8″ and 48° 26′ 52″
7	5	Rule 1: Side opposite = hypotenuse × sine, Rule 2: Side opposite = side adjacent × tangent
	6	Rule 1: Side adjacent = hypotenuse × cosine, Rule 2: Side adjacent = side opposite × cotangent
	7	Page 74
	8	Page 75
	9	After dividing the isosceles triangle into two right-angle triangles
	10	Page 74

Section Number	Number of Question	Answers (Or where information is given in Handbook)
8	I	2 degrees 58 minutes
	2	I degree 47 minutes
	3	2.296 inches as shown by the table on page 964
	4	$\dfrac{360°}{N} - 2\,a$ = angle intercepted by width W. The sine of ½ this angle × ½ B = ½ W; hence, this sine × B = W
	5	3.1247 inches
	6	3.5085 inches
	7	1.7677 inches
	8	75 feet approximately
	9	a = 1.0316 inch; b = 3.5540 inches; c = 2.2845 inches; d = 2.7225 inches
	10	a = 18° 22'. For solution of similar problem, see Example 4 of Section 8
	11	A = 5.8758"; B = 6.0352"; C = 6.2851"; D = 6.4378"; E = 6.1549"; F = 5.8127"; Apply formula on handbook page 79
	12	2° 37' 33"; 5° 15' 6"
	13	5.2805 inches
	14	10 degrees 23 minutes
9	I	84°; 63° 31'; 32° 29'
	2	B = 29°; b = 3.222 feet; c = 6.355 feet; area = 10.013 square feet
	3	C = 22°; b = 2.33 inches; c = 1.358 inches; area = 1.396 square inches
	4	A = 120° 10'; a = 0.445 foot; c = 0.211 foot; area = 0.027 square feet
	5	The area of a triangle equals one-half the product of two of its sides multiplied by the sine of the angle between them. The area of a triangle may also be found by taking one-half of the product of the base and the altitude

Section Number	Number of Question	Answers (Or where information is given in Handbook)
10	1	Page 898 for Morse Page 902 for Jarno Page 911 for milling machine Page 1445 for taper pins
	2	2.205 inches; 12.694 inches
	3	4.815 inches. Page 689
	4	1.289 inches. Page 690
	5	3.110 inches. Page 689
	6	0.0187 inch
	7	0.2796 inch
	8	1.000 inch
	9	26 degrees 7 minutes
11	1	Pages 625, 627
	2	Page 624
	3	Page 624
	4	Page 623
	5	Page 420
	6	Page 421
	7	Page 658
	8	Page 1479
	9	Pages 1474, 1479
	10	When the tolerance is unilateral
	11	See page 633
	12	It means that a tolerance of 0.0004 to 0.0012 inch could normally be worked to. See table on page 632
	13	Yes. See page 704.
12	1	±0.0015, page 424
	2	4000 pounds, page 368
	3	Pages 372 and 373
	4	430 balls, page 2199
	5	¼ inch, page 2236
	6	0.172 inch, page 1473
	7	0.1251 to 0.1252, page 1441
	8	24,000 rpm, page 1457
	9	0.128 inch, page 1418

Section Number	Number of Question	Answers (Or where information is given in Handbook)
	1	Both countries have used the Unified Standard, but England is changing to the ISO Metric. See pages 1474 and 1593.
	2	This is the symbol that is used to specify an American Standard screw thread 3 inches in diameter, 4 threads per inch or the coarse series, and Class 2 fit.
	3	An Acme thread is stronger, easier to cut with a die, and more readily engaged by a split nut used with a leadscrew.
	4	The Stub Acme form of thead is preferred for those applications where a coarse thread of shallow depth is required
	5	See tables, pages 1484, 1485
	6	¾ inch per foot measured on the diameter—American and British standards
	7	Page 1615
13	8	Center line of tool is set square to axis of screw thread
	9	Present practice is to set center line of tool square to axis of pipe
	10	See formulas for F_{rn} and F_{rs}, page 1559
	11	By three-wire method or by use of special micrometers; see pages 1622 to 1640
	12	Two quantities connected by a multiplication sign are the same as if enclosed by parentheses; see instructions about order of operations, page 25
	13	(a) Lead of double thread equals twice the pitch; (b) lead of triple thread equals three times the pitch; see page 1622
	14	See pages 1480–1482
	15	0.8337 inch; see page 1630
	16	No. Bulk of production is made to American Standard dimensions given in Handbook

Section Number	Number of Question	Answers (Or where information is given in Handbook)
13	17	This standard has been superseded by the American Standard
	18	Most machine screws (about 80% of the production) have the coarse series of pitches
	19	(a) Length includes head; (b) length does not include head
	20	No. 25; see table, page 1661
	21	0.1935 inch; see table, page 821
	22	Yes. The diameters decrease as the numbers increase
	23	The numbered sizes range in diameter from 0.0059 to 0.228 inch, and the letter sizes from 0.234 to 0.413 inch; see pages 818 to 824
	24	A thread of ¾ standard depth has sufficient strength, and tap breakage is reduced
	25	(a) and (b) the American Standard Unified form
	26	Cap-screws are made in the same pitches as the Coarse-, Fine-, and 8-thread series of the American Standard, Class 2A
	27	For thread form, see page 1605. There are eleven standard diameters as shown on page 1605.
	28	Page 863
	29	Page 863
	30	0.90 × pitch; see pages 1625, 1626
	31	To reduce errors in the finished thread
	32	Included angle is 82° in each
14	1	A foot-pound in mechanics is a unit of work and is the work equivalent to raising one pound 1 foot high
	2	1000 foot-pounds

Section Number	Number of Question	Answers (Or where information is given in Handbook)
	3	Only as an average value; see page 172
	4	28 foot-pounds; see pages 169 and 170
	5	1346 pounds
	6	Neglecting air resistance, the muzzle velocity is the same as the velocity with which the projectile strikes the ground; see page 165
	7	See page 134
	8	Square
	9	1843 pounds approximately
	10	The pull would have been increased from 1843 pounds to about 2617 pounds; see page 146
	11	Yes
	12	About 11 degrees
	13	The angle of repose
	14	The coefficient of friction equals the tangent of the angle of repose
14	15	32.16 feet per second per second
	16	No. 32.16 feet per second is the value at sea level at a latitude of about 40 degrees, but this figure is commonly used—see page 131
	17	No. The rim stress is independent of the diameter and depends upon the velocity; see page 191
	18	10 to 13; see page 194
	19	No. The increase in stress is proportional to the square of the rim velocity
	20	110 feet per second or approximately 1 mile per minute
	21	Because the strength of wood is greater in proportion to its weight than cast iron
	22	See page 198
	23	In radians per second
	24	A radian equals the angle subtended by

Section Number	Number of Question	Answers (Or where information is given in Handbook)
14		the arc of a circle when the length of the arc equals the radius of the circle; this angle is 57.3 degrees nearly
	25	Page 128
	26	60 degrees; 72 degrees; 360 degrees
	27	Page 127
	28	Page 201 (see page 200 for example illustrating method of using table)
	29	Length of arc = radians × radius. As radius = 1 in table of segments, l = radians
	30	40 degrees, 37.5 minutes
	31	176 radians per second; 1680.7 revolutions per minute
	32	1.5705 inches
	33	27.225 inches
15	1	Page 205
	2	12,000 pounds
	3	1 inch
	4	Page 514
	5	Page 204
	6	Page 514
	7	3-inch diameter; see page 291
16	1	1.568 (see formula on Handbook page 305)
	2	6200 pounds per square inch approximately
	3	It depends upon the class of service; see page 311
	4	Tangential load = 550 pounds; twisting moment, 4400 inch-pounds
	5	See formulas on page 306 and also the table on the same page
	6	The head is useful for withdrawing the key, especially when it is not possible

Section Number	Number of Question	Answers (Or where information is given in Handbook)
16		to drive against the inner end; see page 2235
	7	Key is segment-shaped and fits into circular keyseat; see pages 2241, 2242
	8	These keys are inexpensive to make from round bar stock, and keyseats are easily formed by milling
	9	0.211 inch; see table, page 2240
17	1	See text and footnote on page 2039
	2	American Standard B92.1, page 2017 and 2023
	3	See text, page 2017
	4	See text, page 2017
	5	See definitions, page 2019
	6	None. See text, page 2023
	7	Yes, a crowned spline permits small amounts of misalignment. See page 2036
	8	The torque capacity of splines may be calculated using the formulas and charts on pages 2031 to 2036
	9	Page 2029
	10	The fillet radius permits heavier loading and effects greater fatigue resistance than flat roots through absence of stress raisers.
18	1	18 teeth; 3 inches; 0.2618 inch
	2	2.666 inches; 2.333 inches; 0.1666 inch
	3	Pages 1775 and 1776
	4	Chordal thickness at intersections of pitch circle with sides of tooth
	5	Table, page 1774
	6	Table, page 1779
	7	Surface durability stress and tooth fillet tensile stress are the two principal fac-

Section Number	Number of Question	Answers (Or where information is given in Handbook)
		tors to be found in determining the power transmitting capacity of spur gears. Formulas and procedures will be found on pages 1812 through 1817
	8	Because the tooth shape varies as the number of teeth is changed
	9	No; one hob may be used for all tooth numbers, and the same applies to any generating process
	10	Stub
	11	Page 1777 under Fellows Stub Tooth
	12	Page 1943
	13	Page 1790
	14	Page 1780
	15	See table on page 1776
	16	Pages 1676, 1679 to 1702
	17	Pages 1830 to 1833
	18	Page 1737
18	19	Pages 1843, 1848
	20	Yes, but accurate tooth form is obtained only by a generating process
	21	See paragraph on page 1871
	22	14 teeth; see page 1843
	23	Yes, 13. See pages 1843 and 1847 and compare the ratios and tooth numbers
	24	Page 1842
	25	See text on face angles on page 1843
	26	When the number of teeth in both the pinion and the gear are the same, the pitch angle being 45 degrees for each.
	27	The whole depth minus the clearance between the bottom of a tooth space and the end of a mating tooth = the working depth
	28	See page 1786
	29	See Gear Selection, page 1980

Section Number	Number of Question	Answers (Or where information is given in Handbook)
18	30	See pages 1769 and 1772
	31	See diagram, page 1842
	32	Circular pitch of gear equals linear pitch of worm
	33	Pitch diameter; see Rule number 12, page 1887
	34	Helix angle or *lead* angle of worm is measured from a plane perpendicular to the axis; helix angle of a helical gear is measured from the axis
	35	These terms have the same meaning in each case
	36	Page 1893
	37	To provide a grinding allowance and to increase hob life over repeated sharpenings. See page 1903
	38	Page 1903
	39	See explanation beginning page 1886
	40	Page 1889
	41	Page 1907
	42	Normal diametral pitch is commonly used
	43	Yes (see page 1907), but the hobbing process is generally applied
	44	Pitch diameter
	45	Page 1933
19	1	AISI 1108 CD :½″ = 1008 rpm 12L13, 150–200 HB : = 1192 rpm 1040, HR : = 611 rpm 1040, 375–425 HB : = 214 rpm 41L40, 200–250 HB : = 718 rpm 4140, HR : = 611 rpm O2, Tool Steel : = 535 rpm M2, Tool Steel : = 497 rpm AISI 1108CD :4″ = 126 rpm 12L13, 150–200 HB : = 149 rpm 1040, HR : = 76 rpm

Section Number	Number of Question	Answers (Or where information is given in Handbook)

19	2	1040, 375–425HB : = 27 rpm
		41L40, 200–250 HB : = 90 rpm
		4140, HR : = 76 rpm
		O2, Tool Steel : = 67 rpm
		M2, Tool Steel : = 62 rpm
		AISI 1330, 200 HB : 136 rpm
		201 Stainless Steel, CD : 131 rpm
		ASTM Class 50 Gray
		  Cast Iron : 112 rpm
		6Al-4V Titanium Alloy : 80 rpm
		Waspaloy : 38 rpm
		                (V = 60 fpm)

	3	$1\frac{1}{2}''$ Diam. : 157 fpm—OK 3" Diam. : 314 fpm—OK 4" Diam. : 410 fpm—Too Fast
	4	764 rpm
	5	840 rpm (V = 110 fpm)

	6	Operation
		  1: N = 167 rpm; f_m = 13 in./min
		  2: N = 458 rpm; f_m = 1.8 in./min
		  3: N = 744 rpm
		  4: N = 458 rpm

	7	Existing operation:
		  V = 59 fpm (Too Fast)
		  f_t = .011 in./tooth (Too severe)
		Change to:
		  V = 40 fpm; N = 50 rpm
		  f_t = .006 in./tooth; f_m = 3.6 in./min

	8	Existing operation:
		  V = 188 fpm (Too slow)
		  f_t = .006 in./tooth (Too slow)
		Change to:
		  V = 375 fpm; N = 120 rpm
		  f_t = .012 in./tooth; f_m = 20 in./min

	9	P_c = 7.16 hp P_m = 8.95 hp
	10	V = 98 fpm

Section Number	Number of Question	Answers (Or where information is given in Handbook)
19	11	$V = 205$ fpm ($Q_{max} = 8.55$ in.^{3}/min; $f_m = 10.5$ in./min $N = 131$ rpm)
	12	¼ in.: $T = 123$ lb; $M = 6.38$ in. lb; $P_m = .19$ hp ½ in: $T = 574$ lb; $M = 68$ in. lb; $P_m = 1.0$ hp 1 in.: $T = 2712$ lb; $M = 711$ in. lb; $P_m = 5.3$ hp 19 mm: $T = 7244$N; $M = 37.12$ N · m; $P_m = 2.43$ kW
	13	$T = 1473$ lb; $M = 655$ in. lb; $P_m = 4.9$ hp
	14	Page 966
	15	Page 967
	16	Page 970
	17	Page 970
	18	Page 971 and 972
	19	Page 977, 978, 980, and 981
	20	Page 985
	21	Pages 987 and 989
	22	Page 989
	23	Page 1010
	24	Page 1010, 1012, and 1013
	25	Page 1015
20	1	Drill press, jig-borer, turret punch press, spot welder, riveting machine, shear, inspection machine
	2	Contour milling machine, lathe, grinder, vertical mill, flame cutting machine
	3	NC machines are more productive, more accurate and produce less scrap. See Handbook page 1118
	4	(b)

Section Number	Number of Question	Answers (Or where information is given in Handbook)
	5	Resolver. See Handbook page 1145
	6	CNC systems are less costly, more reliable, and have greater capability than hardwire. See Handbook page 1146
	7	They provide data of slide position and velocity. See Handbook page 1145
	8	At 1.8 degrees per pulse, 200 pulses would be needed to turn the leadscrew 360 degrees, or one revolution. With a 5 pitch screw, the linear movement of the slide would be 0.200 inch, or 0.001 inch per pulse. With 254 pulses, the slide would move 0.254 inch.
	9	True. See Handbook page 1141
	10	(b). See Handbook page 1151
	11	Sequence number. See Handbook page 1158
	12	F. See Handbook page 1151
20	13	Cutter offset is an adjustment parallel to one of the axes. (See Handbook page 1162.) Cutter compensation is an adjustment which is normal to the part, whether or not the adjustment is parallel to an axis. See Handbook page 1146
	14	Four. See Handbook page 1158
	15	False. Circular interpolation eliminates the need for approximating straight lines. See Handbook page 1143
	16	One. See Handbook page 1143
	17	a–5, b–6, c–3, d–2, e–4, f–1. See Handbook pages 1147–1151
	18	False. See Handbook page 1155
	19	False. See Handbook page 1145
	20	(c) See Handbook page 1142
	21	Drive, part and check surfaces. See Handbook pages 1127–1133

Section Number	Number of Question	Answers (Or where information is given in Handbook)
20	22	Drive surface and check surface. See Handbook page 1129
	23	A G word is a preparatory code word consisting of the letter address G, and two digits, that is used to tell the control system to accept the remainder of the block in the required way. See Handbook pages 1156 and 1157
	24	A start-up statement consists of code instructions which will move the workpiece into contact with one or more of the three guiding surfaces (drive, part and check). See Handbook page 1131
	25	The "right-hand rule" says that if a right hand is laid palm up on the table of a vertical milling machine, the thumb will point in the positive X direction, the forefinger in the positive Y direction, and the erect middle finger in the positive Z direction. See Handbook page 1137
21	1	On page 72 is given the formula for length of side S in terms of the given area A
	2	The diameter of each end and the length of taper; see explanation on page 897; also table, 902
	3	Tolerance is applied in whatever direction is likely to be the least harmful; see page 624
	4	It is said that James Watt found, by experiment, that an average carthorse can develop 22,000 foot-pounds per minute, and added 50 per cent to insure good measure to purchasers of his engines ($22,000 \times 1.50 = 33,000$)
	5	Tin in the higher grades, and lead in the lower grades

Section Number	Number of Question	Answers (Or where information is given in Handbook)
	6	Same depth as an ordinary gear of 10 diametral pitch
	7	The tooth thickness and the number of teeth is the same as an ordinary gear of 8 diametral pitch
	8	Add 2 to the number of teeth and divide by the outside diameter
	9	Multiply the outside diameter by 3.1416 and divide the product by the number of teeth plus 2
	10	Birmingham or Stub's iron wire gage is used for seamless steel, brass, copper, and aluminum tubing
	11	Iron wire rope has the least strength of all wire rope materials. See page 365
	12	If surfaces are well lubricated, the friction is almost independent of the pressure, but if the surfaces are unlubricated, the friction is directly proportional to the normal pressure excepting for the higher pressures
21	13	It depends very largely upon temperature. See "lubricated surfaces," page 2201
	14	$8 - (-4) = 12$. See rules for positive and negative numbers, page 15
	15	Yes. One meter equals 3.2808 feet; see other equivalents on page 2424
	16	Experiments have shown that there is definite relationship between heat and work and that one British thermal unit equals 778 foot-pounds. To change 1 pound of water at 212 deg. F. into steam at that temperature requires about 966 British thermal units or 966 × 778 = 751,600 foot-pounds nearly; hence the number of pounds of water evaporated at 212 deg. F. equivalent to

Section Number	Number of Question	Answers (Or where information is given in Handbook)
21		1 horsepower-hour = 1,980,000 ÷ 751,600 = 2.64 pounds of water as given in the Handbook, page 2440
	17	No. The thickness of the pipe wall is increased by reducing the inside diameter; compare thickness in table on page 2378
	18	As a general rule smoother finishes are required for harder materials, for high loads, and for high speeds. (See page 2083.)
	19	Yes. The so-called "O.D. pipe" begins, usually, with the 14-inch size
	20	It is light in weight and resists deterioration from corrosive or caustic fluids. See page 2389
	21	Yes. About 140 degrees lower; see page 2404
	22	Low-carbon alloy steels of high-chromium content; see page 431
	23	A low-carbon steel containing 0.20% carbon or less and usually from 0.60 to 0.90% manganese; see page 450
	24	No. The nominal length of a file indicates the distance from the point to the "heel" and does not include the tang
	25	Yes. See table page 398
	26	Yes. Certain compositions of bismuth, lead, tin, and cadmium; see page 399
	27	(a) and (b) A number indicating how a given volume of the material or liquid compares in weight with an equal volume of water. (c) A number indicating a comparison in weight with an equal volume of air; see pages 395 and 396
	28	The first digit identifies the alloy type;

Section Number	Number of Question	Answers (Or where information is given in Handbook)
		the second, the impurity control; etc. See page 594
	29	Red brass contains 84 to 86% copper, about 5% tin, 5% lead, and 5% zinc, whereas yellow brass contains 62 to 67% copper, about 30% zinc, 1.5 to 3.5% lead and not over 1% tin; see UNS Designations on pages 564, 567, and 568
	30	See pages 2369 and 2370
	31	No. Twenty-six sizes ranging from 0.234 to 0.413 inch are indicated by capital letters of the alphabet (see table, pages 822 to 824). Fractional sizes are also listed in manufacturers' catalogues beginning either at $\frac{1}{32}$ inch, $\frac{1}{16}$ inch, or $\frac{1}{8}$ inch, the smallest size varying with different firms
21	32	To insure uniform heating at a given temperature and protect the steel against oxidization; see page 465
	33	Hardening temperatures vary for different steels; see critical temperatures and how they are determined, pages 523 and 524
	34	Set the taper attachment to an angle the cosine of which equals $0.125 \div 0.1255$; see page 1715
	35	See page 672
	36	Divide $\frac{3}{4}$ by 12; multiply the taper per inch thus found by 5 and subtract the result from the large diameter; see rules for figuring tapers, page 686
	37	Yes. See "Useful Relationships Among Angles" page 81
	38	0.88333; see page 128
	39	$x = 3$

Section Number	Number of Question	Answers (Or where information is given in Handbook)
	40	About 12½ degrees; see page 2222
	41	Ratio between resistance to the motion of a body due to friction, and the perpendicular pressure between the sliding and fixed surfaces; see formula, page 2201
	42	No. Stub's steel wire gage applies to tool steel rod and wire, and the most important applications of Stub's iron wire gage (also known as Birmingham) are to seamless tubing, steel strips and telephone and telegraph wire
	43	If the difference between the lengths of the pawls equals one-half of the pitch of the ratchet wheel teeth, the practical effect is that of reducing the pitch one-half; see ratchet gearing, page 1941
21	44	Apply belt with hair or "grain" side next to pulleys and driving side below so that the slack side above increases arc of contact and pulling power
	45	For a quarter-turn drive, align the center of the driven pulley face with that face of the driving pulley from which the belt leaves; see Angular Drive, page 1266
	46	The ultimate strength is less due to bending action; see formula, page 384 and also table page 388, "Approximate Breaking Strain."
	47	Refer to Handbook page 385.
	48	Multiply 90 by 12 and divide by the circumference of the shaft to obtain R.P.M.; see cutting speed calculations, pages 973 and 974
	49	(a) Lard oil; (b) gasoline

Section Number	Number of Question	Answers (Or where information is given in Handbook)
21	50	If the teeth advance around the gear to the right, as viewed from one end, the gear is right-handed; and, if they advance to the left, it is a left-hand gear; see illustrations, page 1908
	51	No. They may be of opposite hand depending upon the helix angle; see pages 1909 and 1910
	52	Multiply the total length by the weight per foot for plain end and coupled pipe, given in table, page 2378
	53	The processes are similar but the term "pack-hardening" usually is applied to the casehardening of tool steel; see pages 535 and 536
	54	A gas process of surface hardening; see page 538
	55	See definitions for these terms given on page 1244
	56	About 34 inches but the height may vary from 32 to 36 inches for heavy and light assembling, respectively
	57	Major diameter is the same as outside diameter and minor diameter is the same as root diameter; see definitions, page 1477
	58	The S.A.E. Standard conforms, in general, with the Unified and American Standard Screw Thread Series as revised in 1959 and may, therefore, be considered to be the same for all purposes
	59	See information on work materials, page 965
	60	Yes. See pages 510 and 563
	61	13.097 millimeters. See the table on page

Section Number	Number of Question	Answers (Or where information is given in Handbook)
21		2430 which gives millimeter equivalents of inch fractions, inches and feet
	62	The sevolute of an angle is obtained by subtracting the involute of the angle from the secant of that angle. See page 81. The involute and secant functions of an angle are found in the tables beginning on page 82
	63	According to the table on page 975, the recommended cutting speed is 300 feet per minute. This speed is for average conditions and is intended as a starting point, so it is important to know the factors which affect cutting speed as covered in the "Use of Cutting Speed Tables" section on page 971
	64	No. First determine the diametral pitch the same as for a spur gear; then divide this "real diametral pitch" by the cosine of the helix angle to obtain the "normal diametral pitch" which is the pitch of the cutter; see page 1702
	65	Use of alloy steels and reduced hardness in the range of 50 to 56 Rockwell C. See Handbook page 1999
	66	Not in every instance. See page 2040
	67	A cemented carbide seat provides a flat bearing surface and a positive-, negative-, or neutral-rake orientation to the tool insert. See page 726
	68	No. The size of the gear blank, the pitch of the teeth and depth of cut are sufficient for the man in the shop. The tooth curvature is the result of the gear-cutting process. Tooth curves on the working drawing are of no practical value

Section Number	Number of Question	Answers (Or where information is given in Handbook)
	69	By changing the inclination of the dividing head spindle. See page 2074
	70	See formula and example on page 2074
	71	Divide the total number of teeth in both gears by twice the diametral pitch to obtain the theoretical center-to-center distance. (See formula in the table of Formulas for Dimensions of Standard Spur Gears, page 1771)
	72	Subtract number of teeth on pinion from number of teeth on gear and divide the remainder by two times the diametral pitch. (See Rule at bottom of page 1837)
	73	See page 1260
	74	The Standard Wire Gage (S.W.G.), also known as the Imperial Wire Gage and as the English Legal Standard, is used in England for all wires
21	75	A simple type of apparatus for measuring power
	76	With a dynamometer, the actual amount of power delivered may be determined; that is, the power input minus losses. See page 2232
	77	The uniformly loaded beam has double the load capacity of a beam loaded at the center only; see formulas, page 260
	78	Refer to Handbook page 510 for graph of SAE-determined relationships.
	79	No. The nominal size of steel pipe, excepting sizes above 12 inches, is approximately equal to the inside diameter; see tables, pages 2378 and 2379
	80	0.357 inch. See formula, page 225
	81	The laws of sines and of cosines are stated on Handbook page 75
	82	Both the sine and the cosine of 45 deg. are 0.70711

Section Number	Number of Question	Answers (Or where information is given in Handbook)
21	83	Multiply depth in feet by 0.4335
	84	No. See page 2184
	85	100 per cent
	86	50 per cent
	87	Various steels are used, depending on kind of spring. See page 448
	88	This is a special annealing process. The steel is heated above the critical range and allowed to cool in still air at ordinary temperature, page 534. Normalizing temperatures for steels are given on pages 542 to 545
	89	The standard mounting dimensions, frame sizes, horsepower and speed ratings. See section beginning on page 2334
	90	Yes. The American standard drafting-room practice includes section lining, etc. See page 616
	91	No. There are different tapers per foot ranging from 0.5986 to 0.6315 inch; see table, page 898
	92	Yes. See page 1668
	93	Unilateral and plus; see page 624
	94	See table, page 1565
	95	If D = diameter of hole in inches; T = stock thickness; shearing strength of steel = 51,000 pounds per square inch, then tonnage for punching = $\dfrac{51,000\ D\pi T}{2000} = 80\ DT$
	96	See page 415
	97	The Brown & Sharpe or American wire gage is used in each case; see pages 413 to 415
	98	No, this name is applied to several compositions which vary widely

Section Number	Number of Question	Answers (Or where information is given in Handbook)
	99	Antimony and copper
	100	177 nearly; see table beginning on page 1025
	101	See page 398
	102	See page 1029
	103	See page 1032
	104	See page 1171
	105	See page 1715
	106	See page 798
	107	Steel, chromium-plated steel, chromium carbide, tungsten carbide and other materials. See page 713
	108	See text on page 713
	109	The lead of a milling machine equals lead of helix or spiral milled when gears of equal size are placed on feed screw and wormgear stud; see rule for finding lead on page 1716
21	110	Multiply product of driven gears by lead of machine and divide by product of driving gears. If lead of machine is 10, divide 10 times product of driven gears by product of drivers
	111	$\frac{5}{11}$; $\frac{79}{183}$; $\frac{19}{29}$
	112	The whole depth and tooth thickness at the large ends of the bevel gear teeth are the same as the whole depth and thickness of spur gear teeth of the same pitch
	113	See text on page 1246
	114	5.7075 inches
	115	Use the formula (Handbook page 384) for finding the breaking load, which in this case is taken as three times the actual load. Transposing, $D =$ $$\sqrt{\frac{6 \times 2000 \times 3}{54,000}} = 0.816 \text{ inch, say } \frac{7}{8}$$

Section Number	Number of Question	Answers (Or where information is given in Handbook)
		inch diameter
	116	Because the direction of the cutter thrust tends to cause the gear to rotate upon the arbor; see page 1928, "Milling the Helical Teeth"
	117	Due to its collapsing action, the tap may be removed rapidly from the hole immediately after the tapping is completed; see page 893
	118	Chiefly when a hole is to be tapped or reamed after drilling; see page 932
	119	See table on page 491
	120	See page 2312
	121	See page 79
	122	See pages 2005, 1865, and 1866
	123	See page 879
	124	See table on page 1072
	125	See table on page 1067, also text on pages 1067 and 1068
21	126	Motor rotation has been standardized by the National Electrical Manufacturers Association. See page 2334
	127	Yes. See last paragraph on page 2215
	128	To solve this problem the helix angle, ϕ, of the thread at the pitch and major diameters must be found, which is accomplished by substituting these diameters (E and D) for the minor diameter (K) in the formula for ϕ. Thus at the pitch diameter: $$\tan \phi = \frac{\text{Lead of thread}}{\pi E} = \frac{.400}{\pi \times .900}$$ $\phi = 8.052° = 8°3'$ $a = a_e + \phi$ $a_e = a - \phi = 19°16' - 8°3' = 11°13'$ At the major diameter: $$\tan \phi = \frac{\text{Lead of thread}}{\pi D} = \frac{.400}{\pi \times 1.000}$$

Section Number	Number of Question	Answers (Or where information is given in Handbook)
21		$\phi = 7.256° = 7° 15'$ $a_e = a - \phi = 19° 16' = 7° 15' = 12° 1'$
	129	0.0037 inch
	130	$\frac{5}{12}$ foot per minute obtained by dividing 25,000 by 60,000. Note that this speed is considerably less than the maximum surface speed of 150 feet per minute which represents the maximum speed at any load to prevent excess heat and wear
	131	Yes. See Table 13, page 2153 and following tables
	132	550 to 600 Bhn (Brinell hardness number) (see page 2083)
	133	1 pound (see Table 6, page 2199)
	134	See page 2328
	135	23,000 rpm, see page 1458
	136	See page 1634
	137	See footnote, Table 2, page 67

INDEX

PAGE

Absolute efficiency — 111
Acme screw thread, determining normal width — 99
Acme thread tool, checking width of end — 100
Allowance, limit, and tolerance defined — 77
Allowances, for clearance between mating parts — 80
  for "interference of metal" between mating parts — 80
  obtaining by a selection of mating parts — 81
Allowances and tolerances — 76
American and United States standard thread form — 95
Angles, accurate, determining by interpolation — 35
  finding when sine, tangent or some other function is known — 33
  functions — 32
  laying out accurately by use of functions — 36
Answers to "Practice Exercises" and "Review Questions" — 197
Arc, determining radius when length and height of chord are known — 28

Basic size, method of determining — 78
Beam design, examples — 127–130
Beam formulas — 126
Beams, stresses to which they are subject — 125
Bevel gears, checking spherical hub by measuring over plug gage — 53
  dimensions — 157–161
  formed cutter selection — 161
  loads produced on bearings — 166
  power-transmitting capacity — 153
Bolts, calculating tensile strength — 96

Change gear ratio for diametral pitch worms — 165
Chordal dimensions of circles — 4
Chordal thickness, bevel gear teeth — 159

Circles, determining length of tangent when radii and center
   distance are known   45
  dimensions and areas   1
  inscribed in triangle, radius   45
  obtaining given number of equal divisions   4
  segments   5
Coefficient, meaning of   9
Compressive strength calculation   122
Constants   19
Contact ratio, determining   162
Cosine of an angle   32
Cosines, law of   64
Cotangent of an angle   32
Cutting speeds, feeds, and machining power   174

Diametral pitch of gear   153
Drawings, dimensioning to ensure specified tolerances   83
  violations of rules for dimensioning   84
Drilling, estimating thrust, torque, and power for   178

Efficiency of a machine   110
Epicyclic gear sets, worked-out examples involving   152, 153
Equations, trial-and-error method of solution   14

Factor of safety   120
Feeds, cutting speeds, and machining power   174
Foot-pound or pound-foot to indicate turning moment   105
Force, moment of   104
Forming tool angle in plane perpendicular to front surface   46
Formulas   8
  empirical   18
  examples illustrating derivation   15
  omitting multiplication signs   9
  transposition of   10
  trial-and-error method of solution   13
Friction, coefficients of, for screws   113

Gage, angular, checking accuracy by measuring over plug   54
Gears, enlarged pinions to avoid undercut   163

Gears, examples of dimension calculations 154–161
  power-transmitting capacity 153
  power transmitted through gear train 112
  proportioning spur gears when center distance is fixed 156
Gear speeds, calculating 149
Geometrical figures 24
Geometrical principles or propositions, examples illustrating
  application 28, 29
Geometrical propositions and constructions 28

Helical gears, calculating end thrust 164
  helix angle 56
  pitch of hob for cutting 161
Helix angles 56
  helical gear 57
  screw thread 57
Hob, pitch of, for helical gears 161
Horsepower capacity of shafting 132
  determining amount when transmitted through a train of
    gearing 112
  equivalent turning moment or torque 130
  pitch of gearing for transmitting a given amount of power 153

Inch-pounds, turning or torsional moment expressed in 105, 133
Interpolation to find accurate functions and angles 35
Involute splines 141
  calculating dimensions of 141–145
  specifying on drawings 147

Jarno taper, determining included angle 40

Keys, machine, size calculations 135, 136
Keyway formula, depth from top of shaft 38

Limit, tolerance, and allowance defined 77
Link designed to swing equally above and below center line 52
Logarithms, calculations involving 22
  how they are used 22
  of gear ratios 23

Machine tool spindle tapers 74
Machining power, cutting speeds, and feeds 174
Mechanical efficiency 110
Mechanics, miscellaneous problems in 104
Milling, estimating the cutting speed and spindle speed 177
Moments, principle of in mechanics 106
   principle of, compared with principle of work 110
Motor horsepower required when train of gearing is used 112

Negative values in solution of oblique triangles 66
Numerical control 182
   definition 182
   point-to-point programming 183
   continuous-path programming 183

Oblique triangles, formulas for 63
   solution when values are negative 66
   solved by right-angle method 61
   two solutions possible 67

Parentheses 19
Pin measurement of splines 155
Planetary gear sets, worked-out examples involving 152, 153
Pound-foot instead of foot-pound to indicate turning moment 105
Power transmitted through train of gearing 112

Questions, answers to 197
   general review 188

Radians, angles and angular velocity expressed in 115
Radius of arc when chord and height are known 24
Radius of circle inscribed in triangle 55
Rake angle, positioning milling cutter tooth for grinding 42

Screws, coefficients of friction for 113
   force required for elevating or lowering loads with 113
Screw threads, Acme, determining normal width 99
   helix angle 56
   derivation of formula for checking angle 98
   determining pitch diameter 96

pitch diameter, derivation of formula for checking by three-
    wire method   97
Screw thread tools, checking widths of ends   100
Secant of an angle   32
Segments of circles   5
Selective assembly, principle of   81
Shafts, diameter for power transmission   132
    diameter when subjected to combined torsional and bending
        moments   133
    use of keys and set-screws to transmit torque   135, 136, 138
Shearing strength, example of calculation   124
Sine of an angle   32
Sines, law of   63
Slab milling, estimating power required for   176
Speeds, gearing, calculating   149
    geometrical progression   152
Spiral gear, helix angle   57
Splined shaft cutter, determining width at bottom of spline   41
Splines   141
    calculating dimensions of   141–145
    pin measurement of   145
    specifying data on drawings   147
Spur gears, examples of dimension calculations   154, 163
    dimensions for enlarged fine-pitch pinions   163
    proportioning when center distance is fixed   156
Standards, use of   90
Strength of materials   120

Taper hole, checking by means of ball or sphere   51
Taper plug, checking diameter of small end   47
Tapers, examples showing methods of figuring   71
    standards commonly used for machine tool spindles   74
Tangent of an angle   33
Tensile strength, example of calculation   121
Thread milling cutter sharpening position   42
Thread tools, checking widths of ends   100
Tolerances and allowances   76
    dimensioning drawings to ensure specified   83
    limit, and allowance defined   77
Tolerances and allowances, relation to limiting dimensions   76

Tool life   174, 175
Torque or turning moment, how determined   105
   relation to horsepower   130
Trial-and-error, solution of formulas   14
Triangle, radius of circle inscribed in   45
Triangles, oblique, formulas for   63
   solution when values are negative   66
Triangles, oblique, solved by right-angle method   61
   two solutions possible   67
Triangles, right-angle   40
Trigonometrical functions of angles   32
   angles larger than 90°   35
   finding accurate functions by interpolation   35
   finding corresponding angles from tables   33
   laying out angles by means of   36
Turning, estimating spindle speed and power required in   175
Turning moment, for given horsepower   130
   how determined   105
Twisting or torsional moment   105

Unified threads   95
United States standard thread forms   95

Wire measurement of splines   146
Work, principle of, in mechanics   108
   compared with principle of moments   110
Worms, diametral pitch, change gear ratio   165
Wormgear blank dimensions   165